# FINDING A
# SUSTAINABLE BALANCE

# APPLYING GIS

# FINDING A SUSTAINABLE BALANCE

## GIS FOR ENVIRONMENTAL MANAGEMENT

Edited by
**Sunny Fleming**
**Matt Artz**

Esri Press
REDLANDS | CALIFORNIA

Esri Press, 380 New York Street, Redlands, California 92373-8100
Copyright © 2023 Esri
All rights reserved.
Printed in the United States of America.

ISBN: 9781589487581
Library of Congress Control Number: 2023935363

For purchasing and distribution options (both domestic and international), please visit esripress.esri.com.

On the cover: Photograph by Aapsky.

# CONTENTS

# INTRODUCTION

A S WE UNCOVER AND LEARN ABOUT ECOLOGICAL challenges facing our planet, the job of managing environmental and natural resources is becoming increasingly difficult. However, there is room for optimism. Increasingly, the need to address these challenges is being recognized and codified in many ways, including social advocacy, private industry, and public policy. Because of these challenges, there is more demand for applied environmental skill and knowledge, and more opportunity for environmental professionals to apply that knowledge.

Organizations that manage environmental and natural resources face these challenges at a time when they also must address existing needs to prioritize projects that benefit biodiversity; educate, communicate, and engage with communities and stakeholders on environmental issues; and provide transparency on funding and the status and impact of projects. These organizations can benefit from a geographic approach as they address these many needs while operating efficiently.

Incorporating geographic information systems (GIS) into environmental management gives organizations an advantage as they help restore, preserve, and protect our planet. Although populations and habitats grow and change, organizations use GIS technology to manage resources and the environment in sustainable ways—for the long term.

## GIS strategies

Organizations that manage the environment perform many tasks—from regulatory functions to the management of land, wildlife, and recreation. GIS supports mission strategies, including resource management, efficient operations, data-driven performance, and environmental stewardship. These organizations use this technology to create maps that help managers visualize and understand ways to maintain a healthy balance between humans and nature. GIS allows land managers to capture information in the field or remotely. Using GIS, they can analyze, and share insights, and collaborate with partners and stakeholders.

### Land and wildlife management

With GIS, managers can monitor species and ecosystems in real time or take mobile tools offline for monitoring in remote areas. They can use it to capture the extent of management activities and measure their impacts through time and then report on progress and outcomes using spatial analytics and storytelling.

### Outdoor recreation

Outdoor recreation on public lands provides a valuable public service to our communities and supports local economies. GIS offers land managers situational awareness to keep visitors safe. GIS helps them monitor natural resources to inform licensing, interpret resources, and engage with the public through community science. With GIS, land managers can measure and respond quickly to changes in recreational assets, measure and visualize the economic impact of outdoor recreation, and share the status and economic benefits of these assets with stakeholders.

## Environmental regulation

GIS allows organizations to track environmental assets, understand how environmental regulations apply and address potential violations, and prevent environmental hazards from becoming a disaster with early warning. Put simply, GIS provides better insights, faster.

## Stories and strategies

*Finding a Sustainable Balance: GIS for Environmental Management* presents a collection of real-life stories that illustrate how organizations use GIS to solve challenges and create better outcomes. The stories and strategies aim to help the reader understand GIS in the context of The Geographic Approach—the use of geography to help solve problems and make better decisions—for land and wildlife management, outdoor recreation, and environmental regulation. The book concludes with a section about getting started with GIS, which provides ideas, strategies, tools, and suggested actions that organizations can take to build location intelligence into decision-making and operational workflows. The stories and strategies are designed to help readers integrate spatial reasoning into environmental management. The book presents location intelligence as another layer of knowledge that managers and practitioners can add to their existing experience and expertise, offering a geographic perspective that can be incorporated into daily operations and planning.

If the geographic approach isn't currently part of an organization's decision-making processes or considered in daily operational activities, or if it isn't used to improve student outcomes, managers can use this book to start developing skills in those areas. Developing these skills does not require extensive GIS expertise, nor does it require professionals to disregard all their experience and knowledge. The geographic approach adds another way to think about solving problems in the real world.

# HOW TO USE THIS BOOK

THIS BOOK IS DESIGNED TO HELP YOU FOCUS ON ISSUES that matter to you right now. It is a guide for taking first (or additional) steps with GIS. With this technology, you can apply location intelligence to decisions and operational processes to solve common problems and create a more collaborative environment in your organization. You can use this book to identify where maps, spatial analysis, and GIS apps might be helpful in your work and then, as next steps, learn more about those resources.

Learn about additional GIS resources for environmental management by visiting the web page for this book:

go.esri.com/fasb-resources

# PART 1

# LAND AND WILDLIFE MANAGEMENT

ORGANIZATIONS THAT MANAGE ENVIRONMENTAL AND natural resources also aim to protect these resources for a sustainable future. Understanding the relationships between ecosystems and our impacts on them requires a geographic approach. Using the capabilities and tools of GIS, managers can

- monitor species and ecosystems in real time,
- take mobile tools offline for monitoring in remote area,
- capture the extent of management activities,
- measure their impacts through time, and
- use the data to create a conservation plan with spatial analytics.

## Prioritize resource protection

Successful land resource protection begins with setting priorities. With GIS, you can

- assess the status of wildlife resources, distribution, threats, and changes,

- compare scenarios against future modeled conditions to target, and
- plan your stewardship activities accordingly.

## Track stewardship and restoration activities

Resource protection includes stewarding and restoring our natural resources. With GIS, land and wildlife managers can

- monitor assets and resources in the field,
- keep staff safe while executing management plans, and
- track stewardship over space and time to measure progress and outcomes.

## Monitor species and ecosystems

Land and wildlife managers use GIS to

- analyze data from animal collars, cameras, or other sensor networks to remotely monitor species, ecosystems, and environmental variables,
- automatically detect change using imagery and artificial intelligence/machine learning, and
- work in remote areas to efficiently collect observations in the field.

## Educate and engage stakeholders

Education and outreach are components of successful land and wild-life management. GIS allows land and wildlife managers to

- scale impacts through volunteer initiatives,
- collaborate with stakeholders to help inform policy decisions, and

- communicate the success of programs and initiatives to stakeholders.

## GIS in action

This section will look at real-life stories about how environmental management organizations use GIS to protect and manage natural resources. These stories explore how organizations have used the geographic approach and applied geospatial technology to plan conservation initiatives, understand species and habitats, track management activities, and engage with community scientists and other stakeholders to be part of the solution.

# SEEING MULE DEER DECLINE AND CONNECTING THE DOTS FROM ABOVE

### Nevada Department of Wildlife

UNLIKE NEVADA'S LAWMAKERS WHO TEMPORARILY descend on Carson City each legislative session before leaving again, a group of mule deer has taken root. It's an increasingly rare sight to see a herd of this size anywhere in the state and across the western United States, and yet these mule deer have chosen the state capital as their year-round home.

The impact of a harsh winter that killed 30 percent of the state's population of mule deer about 20 years ago is still being felt today, said Cody Schroeder, a big game biologist and mule deer specialist at the Nevada Department of Wildlife (NDOW).

Although a series of winter storms raised the Sierra-Nevada snowpack to a 30-year high in early 2023, drought conditions for several preceding years in arid Nevada have not helped the mule deer either, drying out some of the high-quality vegetation that mule deer need. Other threats to mule deer include grazing competition, invasive species, urban encroachment, changing climate, and healthier predators.

To understand the status of the species and to allocate the number of hunting tags to manage herd sizes, NDOW uses helicopters to tally numbers, gender, age, and health of the mule deer. Until recently, the surveys involved logging data on paper while in the air and often not being able to analyze or visualize the data until much later.

"Now, we can see right away where deer are concentrated and where we have conflict areas," Schroeder said.

NDOW gathers population details and conducts analysis using GIS technology that allows the agency's biologists to understand wildlife and ecosystem health. The data supports decisions to address the ongoing mule deer decline.

## Confronting changing conditions

Advancements in the way NDOW conducts aerial surveys allow for real-time analysis of populations and conditions as NDOW seeks to balance the mule deer's precarious position.

Cody McKee, a biometrician and elk and moose specialist at NDOW, has been working on streamlining data collection, management, and analysis for several years. He was tasked with gathering historical aerial survey data from spreadsheets and filing cabinets and realized that NDOW needed to modernize its workflow for greater efficiency.

A simple big-button form eases input in the challenging data-entry environment of a moving helicopter.

Previously, biologists would take notes and jot GPS points while in the helicopter, and then back in the office they would spend a lot of time typing data into spreadsheets and then merging data and fixing transcriptions errors. NDOW estimates that biologists were spending half the time they spent in flights to get the data usable, so if they spent 1,500 hours flying, it would take an added 750 hours to analyze the data.

"Helicopters are an important part of what we do, for that bird's-eye view of the landscape that gives our biologists a holistic perspective," he said. "It's also a dangerous part of our job, and at least for me, the question 'Is this the last time I get into this ship?' is always in the back of my mind. We need to be sure that we are making the most of our time in the air."

McKee contacted Esri to match the capabilities of ArcGIS® Survey123 and ArcGIS QuickCapture to create one app on one device for the aerial surveying task and save the time it once took to look down instead of forward, where hazards lie. The group

The data of each survey pass is displayed easily on a map alongside the path the helicopter took and what was observed.

worked through iterations to greatly improve what had been a paper-based process, using buttons rather than entry fields to standardize observations.

Using ArcGIS, notes and photos are tagged with position and data goes right to a shared database. The time saved on processing data gives biologists a chance to reflect and study the data to see trends.

For mule deer, and other species, the data supports queries about the cause of decline.

"We used GIS to map the overlap between where mule deer and feral horses are and their preferred habitat," Schroeder said. "We're also looking at other things that are impacting them, such as invasive grass, the drought, and where mountain lions cluster and have kills."

## The cheatgrass problem

One of the most vexing problems that Nevada land managers face is the growing impact of cheatgrass (*Bromus tectorum*), a type of grass that consumes water from native vegetation, and that sprouted in large areas where big fires have burned sagebrush habitat.

"Southern Oregon, southern Idaho, and Utah have had a big problem with cheatgrass, but Nevada has had a cheatgrass invasion," Schroeder said.

In winter, mule deer rely on brush sticking up out of the snow. When cheatgrass is present, native shrubs aren't, so deer have nothing to eat.

The grass flourishes in Nevada's lower arid elevations, where it has altered ecosystems.

"We've documented how cheatgrass changes the natural fire frequencies from 100-to-300-year cycles down to 3-to-5-year cycles," Schroeder said. "It used to be that fires in our ranges were rare, but now they burn over and over."

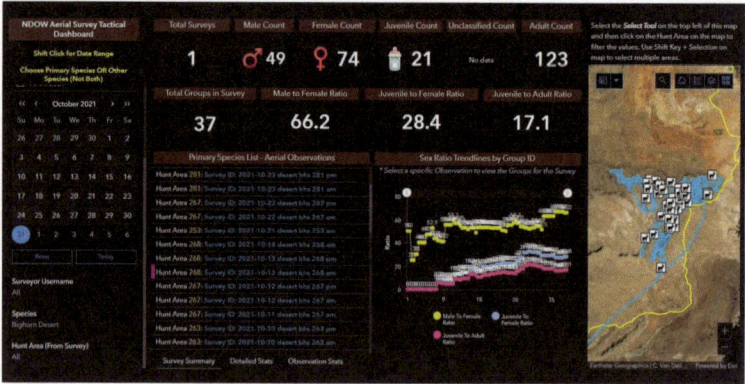

A dashboard view provides details about the current status of survey efforts and the running total of wildlife counted.

## Feral horses multiply, squeezing out other species

More than 80 percent of the land in Nevada is public, and much of it is managed by the Bureau of Land Management (BLM), which is responsible for herd management of burros and feral horses. BLM sets the appropriate management level of herds and works to keep populations down, noting that an appropriate number in Nevada would be 12,811 total. However, it estimates there were 41,853 horses and 4,717 burros as of March 2022 statewide.

Schroeder teamed with David Stoner from Utah State University and other researchers on a paper about the impact of feral horses on other big-game species.

The research involved the spatial analysis of the range of different species such as mule deer, bighorn sheep, elk, and pronghorn antelope overlapped with the range of feral horses. The paper noted that expanding populations of feral horses are a concern for all species.

Researchers are using GIS to study how feral horses impact waterholes during drought and on land with little to no vegetation, and what species are displaced and where.

"The research aims to determine how many horses is too many, and where exactly the conflicts arise around water and forage," Schroeder said.

## Knowing where to apply more management practices

Biologists use aerial surveys to make all manner of wildlife management decisions because of the perspective surveys give them.

"Once you get up into a helicopter, you realize just how connected everything is," McKee said. "While there are many miles separating mountain ranges, the animals we're managing have the ability to cover those miles in a few hours if need be."

With streamlined data collection, NDOW biologists can look at pressures spatially and ask geographic questions from the survey data about population health versus range conditions.

"This is going to help us investigate things and focus our habitat restoration efforts where we can create the most connectivity for wildlife," Schroeder said.

## Addressing regional mule deer decline

In the mid-1990s, the Western Association of Fish and Wildlife Agencies developed a mule deer working group that monitors the population across its full range, addresses disease concerns, and supports best management practices.

NDOW biologists have shared their aerial survey approach with the working group. Peers in all states have the same focus on ensuring longevity of species and making decisions that can sustain populations. With all the pressures mule deer and other species face, this group wants forecasts.

When biologists see drought-stricken rangeland, they know that without adequate precipitation during winter and spring, the wildlife will face challenges in the coming year, McKee said.

Future study of Nevada habitat is planned to guide work in places where conflicts cause the most harm.

Biologists place hope in data-driven collaborations to predict and anticipate further catastrophic change. With the new streamlined workflows providing the ability to compare mule deer reaction to changing conditions, NDOW hopes to engage with other states and stakeholders, including university researchers, to pinpoint causes of decline.

"We don't even know what changes we're going to be looking at in a couple of years," Schroeder said. "But now we can ask these landscape-scale questions."

A version of this story by Mike Bialousz originally titled "Nevada Sees Mule Deer Decline from Above, Connects Dots with GIS" appeared on the *Esri Blog* on January 25, 2022, and was updated in early 2023.

# IMPROVING SAFETY FOR WILDLIFE AND PEOPLE ON ROADWAYS

Center for Large Landscape Conservation, National
Park Service, US Fish and Wildlife Service, and Western
Transportation Institute at Montana State University

WITH ITS WATERFALLS, GEYSERS, AND VARIETY OF WILDLIFE such as bison, cougars and grizzly bears, Yellowstone National Park in Wyoming remains one of the most beloved spots in the United States, drawing nature lovers from around the world to gaze at its wildlife, geothermal wonders, and astonishing vistas.

Every natural feature and species in the park is protected and managed to minimize impacts of visitors who come to experience the awe of nature and wildness. But within and beyond the park's protected federal lands, roads usher millions of visitors and vehicles through wildlife habitats and across migration and movement corridors, creating the potential for dangerous intersections of traffic and animals that threaten motorists and wildlife.

For instance, along the US 191 corridor north of Yellowstone, toward the Big Sky ski area, traffic increased 38 percent from 2010 to 2018, and animal-vehicle collisions accounted for roughly 25 percent of all crashes.

The problem is not isolated. Nationwide, car and truck crashes with animals kill at least 1.5 million large animals annually, and tens of millions of birds and other mammals, including endangered species, are killed each year on US roads. Those crashes also injure 26,000 people and kill about 200 people annually. Additionally, there's up to $8 billion in damages each year.

The need to collect data and map the locations of wildlife-vehicle collisions accelerated with the passage of the Bipartisan

Infrastructure Law in 2021; it includes $350 million to build wild-life road crossings, provided there is data to support and guide the decisions.

Elizabeth Fairbank, who is a road ecologist for the Center for Large Landscape Conservation, said a new tool, known as the Road-kill Observation and Data System (ROaDS), was built using ArcGIS Survey123 to easily gather and analyze spatially accurate data. The ROaDS tool standardizes the information collected in different juris-dictions with spatial accuracy using GPS on a smartphone, she said. "By standardizing the data, you can pool information collected by dif-ferent groups across boundaries that don't exist for wildlife and do meaningful analysis to prioritize the most important places to take action and invest in wildlife crossings at a landscape scale."

## It starts with the data

Researchers plot the collected roadkill observation data on smart maps and conduct analyses using GIS. The National Park Service (NPS) and US Fish and Wildlife Service (USFWS) created the ROaDS tool in partnership with the Western Transportation Institute (WTI) at Montana State University (MSU).

Collecting uniform wildlife collision data nationwide has always been a problem because of the lack of resources and challenge of monitoring nearly 4 million miles of US roads nationwide, 24 hours a day.

The need for action increases as human populations encroach on the natural habitat, vehicle traffic increases, and the climate crisis and wildfires force herds to seek new habitat with food and clean water.

"It's tough on wildlife because beyond protected public lands, we just keep expanding our own footprint of development across the landscape," said Amanda Hardy, a wildlife ecologist with NPS. "Understanding where wildlife live and where they're moving to and from and protecting those areas amid the matrix of development is key to their long-term survival and our coexistence with wildlife."

People can download the ROaDS tool on a mobile device and collect roadkill observation data that can guide mitigation measures needed to reduce these conflicts.

"Identifying wildlife migration corridors where they intersect with our transportation corridors, and where we might have wildlife-vehicle conflicts is going to be really important going forward," Hardy said. "Making sure we put those protections into those areas is absolutely key...but it starts with the data."

## Healthy mix of science, software, and people

The ROaDS tool allows scientists, park rangers, concerned residents, law enforcement, nonprofit groups, and students to collect information that helps them implement solutions that have been elusive because of insufficient data collection.

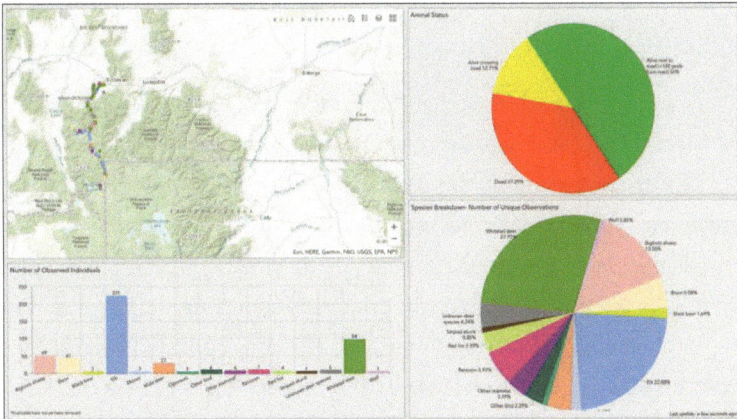

Details collected in the ROaDS tool for US Route 191 in Paradise Valley, Montana, are aggregated on this dashboard to see the wildlife-transportation interactions across space and time.

Two projects around Yellowstone have drawn interest among volunteers who want to help document vehicle-wildlife collisions. Using the ROaDs tool, they can note GPS coordinates, time of observation, date, and wildlife species observed, among other information.

"We have hundreds of wildlife observations recorded so far on the two highways we are monitoring in the Greater Yellowstone Ecosystem: Highway 191 (west of Yellowstone) and Highway 89 in the Paradise Valley (just north of Yellowstone). So, there's been quite a bit of use at least by a handful of dedicated people who are going out and doing this on a regular basis," Fairbank said.

Fairbank said the aim is to provide usable information so planners can select the appropriate type of wildlife crossings and build them in the most important and effective locations. "The goal is to provide information for transportation planning that accommodates safe passage for both wildlife movement and human movement, and that can actually save money in the long run by reducing wildlife-vehicle collisions"

## Wildlife-vehicle collisions keep increasing

The increase of wildlife-vehicle collisions can jeopardize the survival of species that may already be threatened or endangered.

Given that many wildlife-vehicle crashes are not reported to law enforcement or insurers, ROaDS can help clarify the numbers.

The ROaDS tool reflects the focus on endemic species in an area or fits the purpose of a diversity of groups collecting slightly different types of data. WTI plans to compare the data collected by the Confederated Salish and Kootenai Tribes to data collected by the Montana Department of Transportation in the same sections of US Route 93 on the Flathead Indian Reservation, said Mathew Bell, a research associate with WTI.

By taking samples from different groups, it may be possible to produce better estimates.

"Living in Montana, we know to watch out for wildlife," Bell said, "but as you get the public involved in these things, they start to acknowledge how many are actually getting hit, how much wildlife are dying, and the safety concerns."

## Aiding research in diverse places for diverse species

Although the work around Yellowstone National Park tends to focus on larger mammals, a nonprofit group is tracing movements and tracking mortalities of desert tortoises, a threatened species listed under the Endangered Species Act, in the Mojave Desert.

Many smaller animal species are injured or killed in vehicle collisions, potentially upsetting the balance of nature in an area or larger region. Smaller species can also cause crashes as motorists swerve or stop for all sizes of animals.

"We may not be dealing with huge collisions that get a lot of news coverage, but various species in our refuge units are getting hit," said Vincent Ziols, manager of the Transportation Safety Program for the USFWS National Wildlife Refuge System. "It's very difficult to protect the integrity of complex ecological systems when there may be a dozen or so species at risk of becoming roadkill at one given point."

According to the Federal Highway Administration, road mortality is one of the major risks to the survival of 21 federally listed threatened or endangered animal species in the United States.

The ROaDS tool allows observers to attach a photo of the deceased animal to the data-collection form, improving identification and documentation of mortalities of rare or at-risk species.

"ROaDS will go far to highlight the need to protect all wildlife from our nation's roads, and not just in those areas that have already received attention from the public," Ziols said. "Those in conservation can use this tool to address these issues all over the nation for both big and small animal species."

In Canada's Banff National Park, mitigation measures and a series of wildlife crossings, including bridges and tunnels allowing wildlife to cross over and under roads, have reduced collisions between vehicles and all species of wildlife by more than 80 percent.

Data shows that the number of elk and deer hit by vehicles has been reduced by more than 96 percent in the park. The mitigation measures are an example of how a concentrated effort supported by research and funding can increase driver safety and wildlife survival.

In the United States, the $350 million contained in President Biden's Bipartisan Infrastructure Law potentially can fund many safe passage projects for animals in areas with high wildlife road mortality.

"You probably aren't going to get one of those grants unless you come to the table with the information that justifies why you would be putting an investment at a given site," Hardy said. "But it starts with data."

Ultimately, the ROaDS tool enables agencies and citizens to collect the data needed to invest in measures that reduce wildlife-vehicle collisions.

A version of this story by Sunny Fleming originally appeared on the *Esri Blog* on January 4, 2022.

# HUB BUILDS COLLABORATION FOR FOREST PLAN

## Montana Department of Natural Resources and Conservation

MONTANA'S APPROACH TO IMPROVING FOREST HEALTH and reducing wildfire risk statewide includes a web-based location platform in which the state makes forest data accessible. Forest stakeholders used the platform to collaborate on completing the 2020 Montana Forest Action Plan.

Good stewardship and management of forests is essential to many Montanans. Recreationalists; the forest products industry; federal, state, tribal, and local-level land managers; private forest landowners; representatives of conservation organizations; collaborative and watershed groups; ranchers and farmers; wildlife watchers; and other partners all have vested interests in the health of the state's forests. Although these stakeholders value the state's 23 million acres of forested land in different ways, they share the goal of keeping Montana's forests healthy and resilient.

Under the authority of the 2008 and 2014 Farm Bills, Congress tasked states and territories with assessing the condition of the forests within their boundaries, regardless of ownership, and developing strategies to promote forest health and resiliency through a state forest action plan. Montana's first plans were static hard-copy or digital documents that left little room for change or iterative versions.

When it came time to revise the Montana Forest Action Plan in 2020, the Montana Department of Natural Resources and Conservation (DNRC) became the principal agency responsible for the revision of the state's forest action plan. DNRC wanted to use technology that would ensure the plan's continued relevancy.

Montana's web-enabled plan would be a living document—easy

to access, explore, and update. By using ArcGIS Hub℠ applications, the 2020 revision incorporates the most up-to-date data and science as it becomes available, thereby providing accurate and timely information as Montana changes over the next decade.

More importantly, the hub will show, at landscape scales, how the Montana Forest Action Plan has changed forest health and wildfire risk and communicate that information in ways that are accessible and easily understood.

DNRC's GIS team had experience developing web mapping applications, which bolstered revision efforts. The team previously created the Montana Interactive Wildland Fire Information Tool for sharing wildfire information with fire managers, decision-makers, and the public. This web app proved its worth, particularly during the intense 2017 fire season, when thousands of individuals used it daily to see current information about fire conditions.

DNRC GIS manager Brian Collins thought a similar solution would work for the 2020 Montana Forest Action Plan. This time, he wanted to use the GIS platform to build an information center for participants to share their data, ideas, and goals.

Collins knew that buy-in from top department executives, who make decisions and build policies, would be required for the project's success. He set up a meeting with Sonya Germann, Montana's then-state forester and administrator of the DNRC Forestry Division. Collins pitched the idea of implementing a GIS platform to serve as a focal point, which would encourage collaboration and engagement with the Montana Forest Action Plan.

"The 2020 Montana Forest Action Plan is an opportunity to use geospatial technologies that improve planning, reporting, and understanding," Collins said.

Collins described GIS capabilities to Germann by demonstrating the functionality of ArcGIS Dashboards and ArcGIS StoryMaps℠

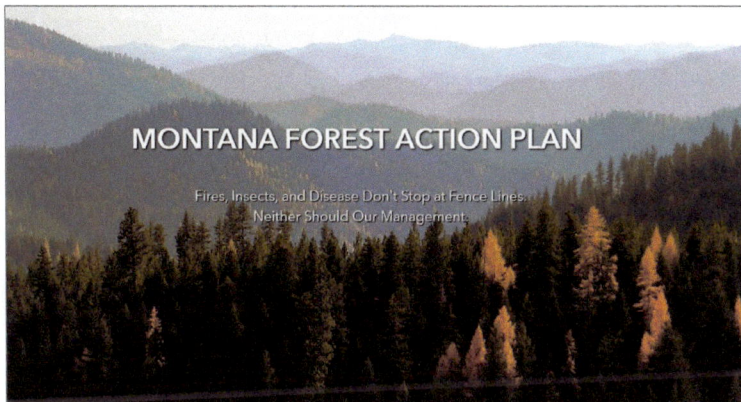

MONTANA FOREST ACTION PLAN

Fires, Insects, and Disease Don't Stop at Fence Lines.
Neither Should Our Management.

Montana Department of Natural Resources and Conservation revised its state's forest action plan using Hub to ensure the web-enabled plan is easy to access, explore, and update.

software. He then introduced the concept of a cloud-based engagement platform with a hub for sharing information between departments and engaging the public. It could be the venue for planning and collaboration.

Hub manages content and data and can display them as maps, dashboards, StoryMaps stories, documents, and website pages. Organizations use hubs to gather data for their projects and contribute their own data for others to use.

During the Montana Forest Action Plan revision process, staff and relevant partners used the hub to collaborate and share information and ideas. The public and important forest stakeholders also used the hub to engage in conversation and submit feedback.

"I explained how ArcGIS Hub consumes different types of data and then represents that information through different means, so it connects with people in ways that make sense to them," said Collins. "When it comes to communicating data, we need to make it as accessible as possible, and ArcGIS Hub allows us to do that."

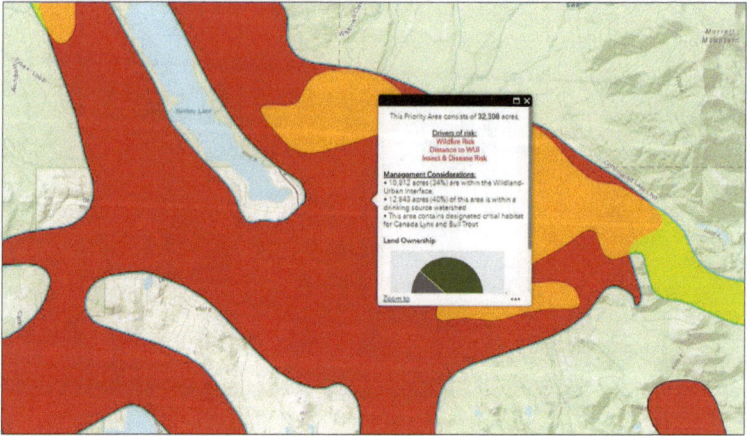

This is an example of one of the dashboards showcased in the web-enabled plan that identifies priority areas with dynamic pop-up content, such as risks and management considerations.

Germann saw value in the technology and supported the project. The GIS platform and hub would support the Montana Forest Action Advisory Council. The council is a group assembled by Montana's governor to help develop the Forest Action Plan and implement strategies to help improve and sustain forests throughout the state. Germann also ensured that the GIS manager played a role on the project's core team.

The DNRC GIS team implemented the Hub platform and used the department's existing open data to make and share maps. Data resources grew because agency partners and council members began to share open data via the hub.

"Hub has enabled us to reduce the friction between data and the people that need to be informed by it," said Nick Youngstrom, lead geospatial analyst for the Forest Action Plan team. "We've removed some of the technical knowledge and staff power needed to use and access geospatial information on Montana's forests."

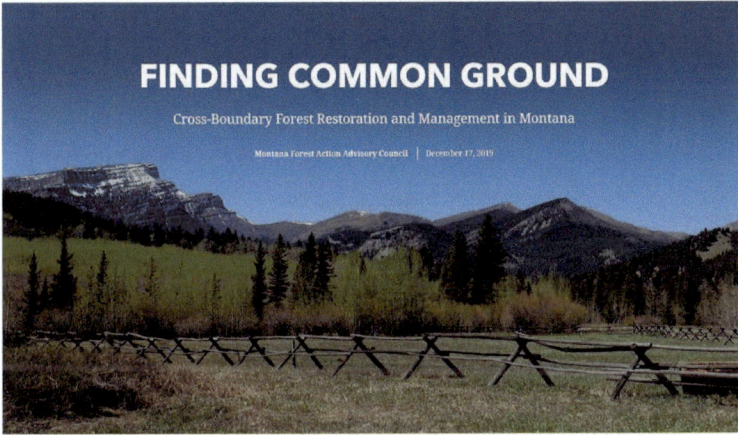

# FINDING COMMON GROUND

Cross-Boundary Forest Restoration and Management in Montana

Montana Forest Action Advisory Council  |  December 17, 2019

The Montana Forest Action Plan used ArcGIS StoryMaps to prepare a story that explains the concept of cross-boundary forest restoration management.

All the council's data, as well as interactive maps and additional information, is available on the Montana Forest Action Plan website (montanaforestactionplan.org). Visitors can explore data layers used in the plan, such as wildfire hazard potential and an interactive analysis of Montana's urban forests. Participants can also dig deeper into the plan's data, using the Hub dashboard to visualize and understand information and track progress toward the accomplishment of goals and objectives set by the council. The GIS team created mapping apps to help the public better understand where projects are located and the specific goals or purposes of those projects.

Additionally, the website hosts a storytelling map that presents the plan's initiatives. For instance, to help illustrate the Shared Stewardship initiative, the *Finding Common Ground* story explains the concept of cross-boundary forest restoration management. Although the topic sounds daunting, the story format makes the concept understandable and engaging.

"Hub technology made this planning process, by far, the most successful effort we've had for collaborating on data collection efforts with a multitude of partners," Collins said.

The Montana Forest Action Plan is a living document that can evolve with the times. Its platform serves as a nexus of information, analysis, and engagement and ensures that all Montanans have the information and resources they need to keep Montana's forests healthy and resilient.

A version of this story originally appeared on esri.com in 2020.

# PLANT BACK BETTER

## American Forests

THE CAMP FIRE, AN ENORMOUS CALIFORNIA WILDFIRE IN 2018, claimed 85 lives and consumed the entire town of Paradise. Ever since, experts have been devising ways to safeguard against another tragedy and rebuild the forest destroyed by the Camp Fire. Rather than simply replant what was there, the Bureau of Land Management (BLM) set out to map a climate-informed restoration plan.

"We want to plant it back better to withstand wildfire and future climate, so the community is not vulnerable like that again," said Coreen Francis, Nevada state forester at the BLM.

During her more than 20-year forestry career, Francis has seen shifts in forest health from drought, insects, disease, and climate. The pace of change in the forests around Paradise forced everyone to reexamine their understanding and try to catch up. To create a smart restoration plan, she convened experts to combine their knowledge about the land and forest using GIS to build a sustainable plan.

In less than a decade, several fires burned across the same area that was devastated by the Camp Fire, which burned 153,336 acres. Since 2018, more megafires include the North Complex Fire that consumed 318,935 acres in 2020 and the Dixie Fire that burned 963,309 acres in summer 2021. Together, these fires left relatively few trees untouched in the burn areas of Northern California.

Because the climate has changed, the types of trees used to replant the area have also changed, according to Austin Rempel, senior manager of reforestation at the nonprofit American Forests. "For instance, sugar pine is everyone's favorite tree because they grow big and look nice, but climate models say they don't want to live here anymore. Low-elevation sugar pine is going to be a thing of the past."

## Assisting tree migration

Trees can't just pick up their roots and move, and a natural migration could take centuries. It's up to foresters to plant for what the forest wants to become, a practice known as "assisted migration."

"Assisted migration is a no-brainer for our organization, knowing that forests need to adapt," Rempel said. "In the Camp Fire area, because of its low elevation, it's quickly turning from dense mixed conifer forest into a place that wants to be more oak and grassland and chaparral and gray pine."

Analysts at American Forests apply models that use spatial analytics to consider species tolerances and soil types, along with climate forecasts about heat and rainfall, to predict what plants will want to live in a place, far into the future.

This level of climate action requires a detailed map to understand what exists, the conditions best suited for each plant, and where similar conditions can be found elsewhere. GIS is used to perform this suitability analysis, with predictions that improve with more data.

For Francis and other foresters, ArcGIS Online became a repository where they could combine data, plan collaboratively, and view a shared map on portable devices as they roamed the burn scar. Checking the map in the field is called ground truthing, and it provides the opportunity to adjust and add more details.

"Some data we had was wrong," Francis said. "Being able to see it right there allows us to build knowledge and make our plan a little more accurate. We take scientific concepts, and we look at them on the ground, and then we compare them with what we see on the map," she said. "We can scroll and look at different layers while we're walking to inform us of things that we can't readily see. Knowing the soil type is serpentine, for example, explains why those trees look scrawnier."

Wolfy Rougle of the Butte County Resource Conservation District surveys replanting plans. (Photo courtesy of Austin Rempel, American Forests)

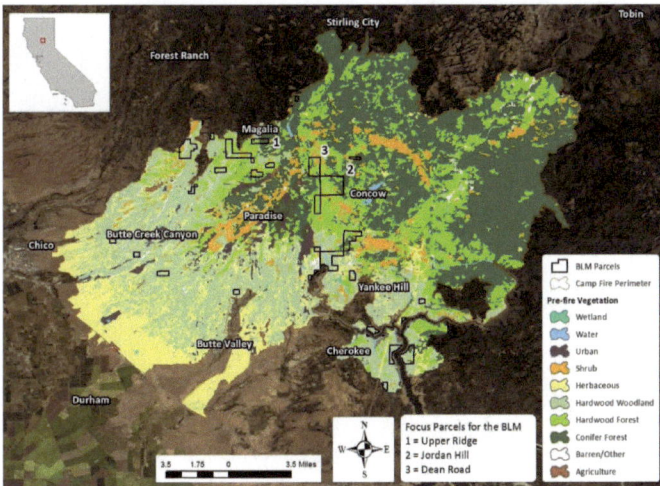

Before the Camp Fire, this mix of vegetation was present in the forest.

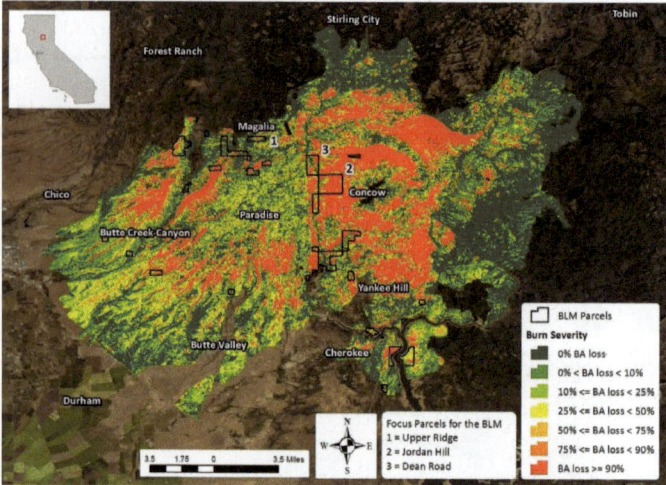

The severity of the Camp Fire, like all fires, was not uniform. Here, red areas denote where more than 90 percent vegetation loss occurred.

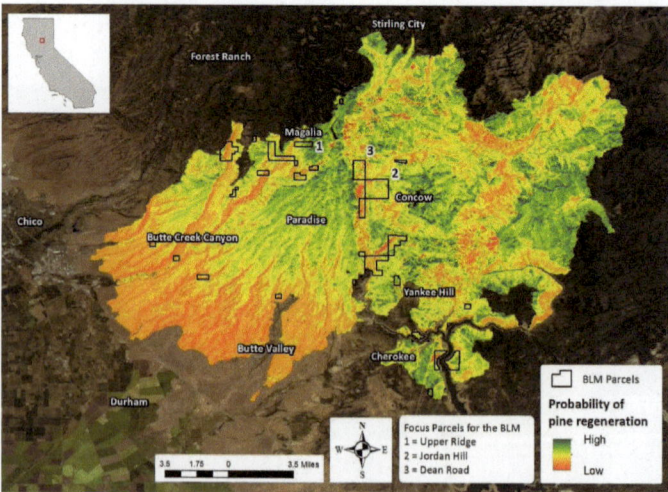

Comparing the climate forecast with suitable conditions for pine trees shows where best to plant them (green areas on the map).

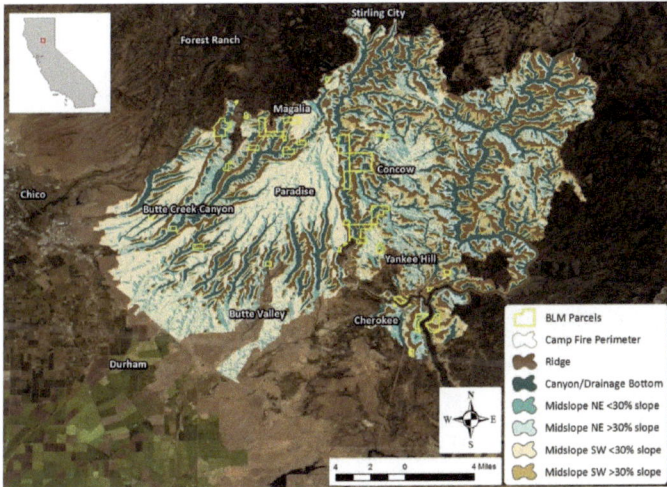

The topography—slopes and lowlands—determines much of the postfire forest management strategies.

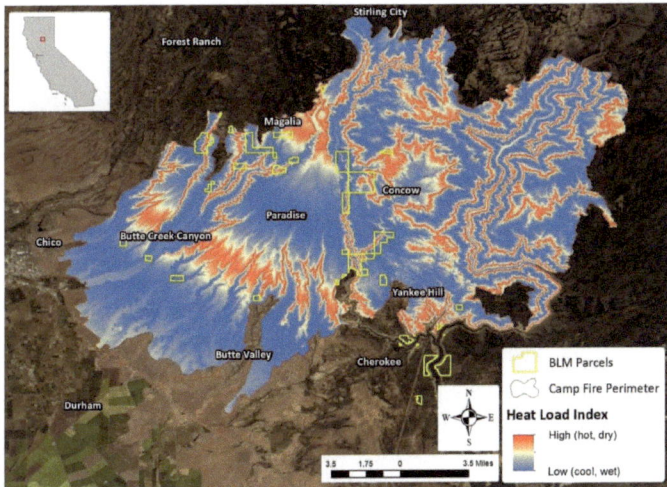

This solar heating map denotes cool and wet versus hot and dry areas to guide planting.

Much of BLM's management practices are guided by shared maps. GIS is well suited for planning at the landscape level because it contains details on the topography—ridges, rock outcroppings, slopes, water, valleys. Foresters must consider multiple factors, among them: north-facing slopes are cooler, south-facing slopes are drier, and valley bottoms have the deepest and best soils.

"Mapping the landscape is a starting point," Rempel said. "It shows us what the forest should look like and what we should plant there."

The map pinpoints the places that will be climate stable and ideal for planting specific species.

"We know where trees live now, and we can model the climates they're comfortable with," Rempel said. "We can use GIS to map the soil productivity and where trees would be most successful."

The model and map include ecology, with data to analyze and explore the pieces of the environment that contribute to a tree's survival. GIS becomes a repository of earth processes and a way to query and model to apply nature-based solutions to restore ecosystem balance.

"We've talked about the concept of island plantings, where you put a diversity of species into a small plot, maybe a quarter of an acre, and grow those in clumps or islands across the landscape," Francis said. "Eventually trees will produce seed and the seed will burst into the surrounding area, and it promotes more diversity on the landscape."

GIS also was used to plan and create natural fire breaks in the landscape to reduce the intensity of future fires.

The map helped speed the reforestation by picking the areas to plant first where they will have the most strategic advantage.

## Common ground for collaboration

Multiple stakeholders and participants were involved in making the climate-informed restoration plan. BLM guided the effort with the help of American Forests and participation from the US Forest Service, CAL Fire, Plumas National Forest, Butte County Fire Safe Council, Sierra Pacific Industries, and others.

Having a timber company at the table is unusual, but so is what happened to Sierra Pacific Industries' part of the forest that burned in 2012. The company diligently replanted it in hopes of harvesting lumber, and then just six short years later, the Camp Fire burned everything the company planted. "That was enough for them to say, 'This is not a place where we can do production forestry anymore,'" Rempel said.

The stakeholders came to the planning sessions with ideas, maps, and open minds. The evidence was clear: everyone was wasting their time by doing the same things that had been tried before.

"Permaculture ideas—nature-based approaches—are starting to enter into forestry," Rempel said. "It takes a very long time to convince old school foresters that this is the way, but it is happening slowly."

ArcGIS Online became the place where everyone could work and iterate together. For those not familiar with GIS, they could view the maps and agree or disagree with what they were presented.

"The sharing platform was central to our collaborative approach and our climate conversations," Rempel said. "We had these sessions during different versions of the draft where we got all the land managers and foresters together to go over what they were seeing or if other tricks of the trade should be added to the report."

## Tackling a trend

The foresters who crafted the Camp Fire restoration plan hope that climate-informed strategies become more common; the approach is practical in making the most of limited resources by pinpointing the places where the forest can thrive.

"Many of the climate plans just offer big-picture ideas—about techniques that could be applied," Francis said. "Our plan takes those large concepts to the ground level. Predictions of what the climate is going to be informs our implementation plan."

According to research at American Forests, 81 percent of reforestation needed on national forest land is now due to wildfires rather than logging.

To replant wisely, new models must factor in future climate.

"This is a recovery plan," Francis said. "It's about using the best science to replant."

A version of this story by Mike Bialousz originally appeared in the Winter 2022 issue of *ArcUser*.

# REAL-TIME TOOL TRANSFORMS WILDFIRE FIELD OPERATIONS

## Pennsylvania Department of Conservation and Natural Resources

WHEN THE SEVEN PINES FIRE IGNITED IN EARLY NOVEMBER 2020, it quickly consumed acres of scrub oak and mountain laurel along steep terrain near Swiftwater, Pennsylvania. Firefighters worked according to COVID-19 distancing needs and used live maps to concentrate containment efforts and protect homes. After four days, and with 812 acres burned, the human-caused blaze was out.

Fire officials managed the blaze in the context of COVID-19 protocols in place at the time, said Shawn Turner, forest fire specialist supervisor in the Pocono Mountains for the Pennsylvania Department of Conservation and Natural Resources (DCNR). "We couldn't have a whole lot of people on the fire, because then we'd have to deal with increased exposure."

The response effort included a regional team of firefighters and equipment such as a bulldozer to scrape the ground for fire breaks, a helicopter for targeted water drops, and reconnaissance aircraft to relay photos of the fire's extent. Firefighters mobilized this response via a live map app called the Fire Mapper app, built with GIS.

The Fire Mapper app let them see the work of others and report their own actions while keeping everyone away from each other because of the pandemic. It helped the teams orchestrate efforts as fire and weather patterns changed moment by moment.

"We had very low humidity for two nights after the fire started," Turner said. "This was in a remote area with dry conditions that hadn't burned since the early 1970s, so the fuel load was high. We knew the fire could get large on us."

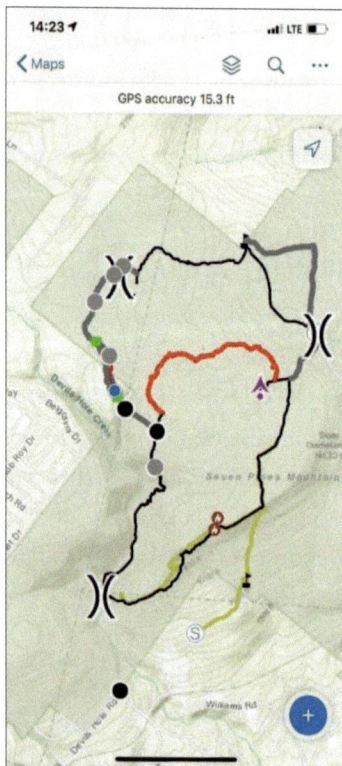

The live Fire Mapper app contains the current location of the fire line (in red) and the containment lines created by firefighters and bulldozers (in black).

If the fire crested a ridge beyond the state game lands where it originated, people in the 1,000 homes of the Pocono Farms East neighborhood would be forced to evacuate.

Fortunately, humidity increased, and the fire advanced in a northern direction, staying below the ridge and moving toward a more mature hardwoods section of forest with fuels that don't burn as readily. "As soon as nighttime humidity recovered, like it normally does, it slowed the fire growth and gave us the time to get it," Turner said.

## The live fire map takes hold

Turner, who functions as an incident commander, has been enthusiastic about using Fire Mapper to guide field operations.

"When I started in 1990, we still used a compass to do surveying for wildfires," Turner said. "We moved to GPS and GIS, but we still waited for a day or more to get a map out to the field. Now people are seeing what we can do with the live map."

The Fire Mapper app was first created during the 2014 fire season by Matt Reed, operations and planning section chief for DCNR Division of Forest Fire Protection, with help from Chad Northcraft, a fire forester turned air tanker base manager. The two shared a vision for a live map that could help teams efficiently fight fires. Reed built an app using ArcGIS Collector technology.

"There are four things that we're constantly aware of in fighting fire that we call LCES—lookouts, communications, escape routes, and safety zones," Reed said. "One of the first things we want to do is see where the fire is at and just pay attention. We always strive to have someone up in the air to relay information through the map."

Fire mapping was used selectively during the first year, but it caught on statewide in 2015. Anyone involved in wildfire or prescribed fire activities within DCNR has access to the app so they can record their work.

"You know the saying, 'a photo is worth a thousand words'? It is," Turner said. "When somebody's describing something to you, sometimes you're not exactly sure what it is they're talking about. But if they can take a photo and attach that to the map, then you get a really good idea of what's happening."

In 2016, dry conditions sparked 850 wildfires in Pennsylvania. In April of that year, the 16-Mile Fire burned 8,000 acres in the Pocono Mountains in northern Pennsylvania, requiring 130 firefighters who finally got it under control after two weeks.

Firefighters start a burnout operation to remove unburnt fuels between the fire and containment lines during the Seven Pines Fire. (Photo credit: Matt Reed)

"The 16-Mile Fire got very complex; it started as two fires that then came together," Reed said. "Fire Mapper got a lot of use and a lot of exposure because we had people from all over the state responding to that one, and even people from outside the state."

## Seeing changing wildfire behaviors

In 2020, fires consumed five million acres across the western states of Washington, Oregon, and California. Wildfires scorched more than 800,000 acres in California alone. The fires have been called climate fires because they reached farther and burned longer due to warmer and drier conditions.

Climate fires also threaten eastern US states, putting the 1.1 million acres of the Pinelands that span eastern Pennsylvania and western New Jersey at high risk. Recent droughts and dead trees from gypsy moth infestations have also heightened concerns for large fires in the Poconos.

Contrary to the Seven Pines Fire, it's not typical for a fire to burn

The new fire tower in Big Poconos State Park replaces a tower that was staffed by fire spotters during spring and fall fire seasons since 1921.

for long or far in the Poconos in November. More than 80 percent of wildfires in Pennsylvania occur in March, April, and May. It's during those months when winds are higher and the sun hits the forest floor and dries out the leaf litter that wildfires can thrive. In 2020, dry conditions created an unusually high number of fires in February—161 compared with 11 a year earlier. Scientists expect ongoing shifting seasonal patterns due to climate change.

The region's devastating 2016 fire season reinforced the need to invest in new fire lookout towers in the Poconos, replacing 16 towers built in the 1920s with new modern structures. While most states are shutting down and reducing fire towers, Pennsylvania has found lookouts to be cost-effective and superior to other monitoring methods in remote areas, so it's building more.

"There are a lot of places where if people see smoke, they will get 10 calls to the 911 center," Mike Kern, chief of the DCNR Division of Forest Fire Protection, told National Public Radio. "There's other places in north-central Pennsylvania especially where no one will see a fire for a couple hours."

The towers are a front-line defense to spot smoke quickly and keep fires small. The live Fire Mapper app collects all inputs and displays data for everybody, including key inputs from lookout towers.

He recently talked to a dispatcher who uses the tool in a district where fire spotters in the towers call in with the bearing and distance of smoke columns, which the dispatcher adds to the map, Reed said. Then using a dashboard, the dispatcher determines where available personnel are at the time and dispatches the closet responders automatically, sending them a location through the map.

Annual reviews help Reed and his team hone the solution after each fire season and tailor it to meet the needs of more users. Despite the live map's strong success over more than five years, some Pennsylvania fire districts have yet to adopt it.

"We seem to be battling the 'that's not the way we've always done it' crowd," said Chad Northcraft, incident management specialist and Mid-State Tanker Base manager at Pennsylvania DCNR. "Anybody that puts their hands on it sees the way we're using it and its capabilities, and they're immediately dialed in and know it's the way to move forward."

## Five years of incremental improvements

Fire Mapper spans many users who each view the map and use it for their own needs while contributing to the shared awareness of others.

"We have a recon plane in the air anytime it's a fire day, and they're up there mapping fires," Northcraft said. "The more time I spend at the tanker base, the more I realize how the tool can be used in different ways."

Incident management teams use Fire Mapper to track the location of firefighting assets, study vulnerable properties and structures,

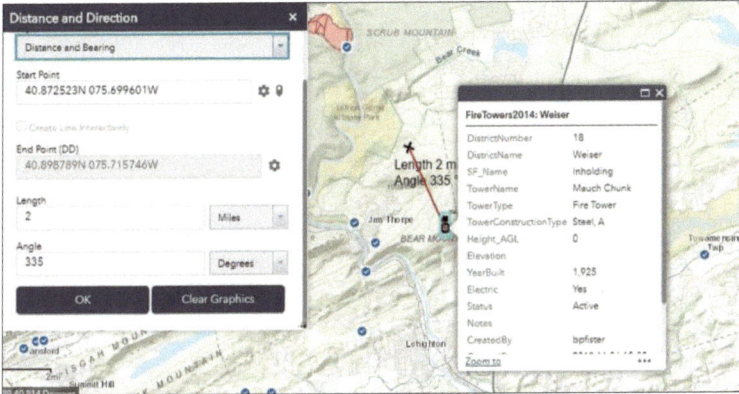

The Distance and Direction web app helps pinpoint the location of a fire when smoke is spotted.

and figure out how to access a fire. Leadership also shares the information to keep the governor and the public informed.

"I was on the Seven Pines Fire, but my supervisor was not," Reed said. "He could just pull this right up on his screen and see what's going on and give his report to the state forester."

"It is important to have quick and accurate information when dealing with dynamic incidents such as wildfires," said Ellen Shultzabarger, Pennsylvania's state forester. "It's beneficial to have this nearly real-time data coming straight from the field for reporting and decision-making needs."

Before calling in resources, response teams put a virtual pin on each fire, determine their directions and how fast they're spreading, and draw polygons around each fire represented on the map.

For the Seven Pines Fire, the response team achieved a new level of awareness because a larger group of responders recorded their actions using a new tracking function.

"I couldn't believe how simple it was to have the dozer boss mark his points," Reed said. "We were able to just convert the dozer line to the controlled fire edge. We had another guy scouting to guide the dozer. It was so simple to say, 'Hey, Greg, can you draw a line in where you're proposing that dozer line?' Greg says, 'Yeah, here you go.' And there it is on the map."

A version of this story by Mike Bialousz originally titled "Real-Time Tool Transforms Wildfire Field Operations in Pennsylvania" appeared on the *Esri Blog* on January 12, 2021.

# INTO THE WEEDS

## Wild + Pine

ROM THE OUTSIDE, IT'S EASY TO THINK OF THE PROVINCE
of Alberta, Canada, as predominantly untouched wilderness. But
Alberta's natural environment has endured its fair share of industrial
land clearing. In response, environmental restoration organizations
like Wild + Pine (W+P) have set out to reverse the impacts of industrial development.

W+P—which manages a 12-hectare (30-acre) site near Pine Lake,
Alberta, on behalf of the Nature Conservancy of Canada—aims to
transition grassy fields that are no longer suitable for conventional
agriculture back to natural forest. After the Pine Lake site was pre-
pared and planted with native seedlings, however, it faced a familiar
and deceptively simple problem—weeds.

Weeds often grow fast and can quickly outcompete seedlings,
resulting in failed reforestation. But managing weeds is laborious and
time-consuming.

"We wanted to be strategic about our approach to weed control," said Jolan Aubry, field supervisor at W+P. "But we didn't want
to spend hours mapping, plotting, and analyzing things; we needed
something efficient."

To that end, W+P began collaborating with TerraLab, an Aus-
tralian technology company that produces the STA logger, a location
and data logging device.

In addition to the Global Navigation Satellite System (GNSS)
device that collects location data, the STA logger also has an attach-
ment that connects to the handle of the spray equipment and monitors its activity. When the spray equipment releases herbicide onto
the weeds, the STA logger captures the event and records its location.

The STA logger attached to a backpack sprayer used for controlling weeds.

Recording weed locations quickly in the field means W+P field operators are effectively doing two jobs at once—mapping and spraying weeds.

## From field to GIS framework

On a typical day at Pine Lake, Aubry and his team can arrive at the site, put on their safety gear, pick up their backpack sprayers, and get to work quickly.

"There are no complications with signing in, starting projects, or even turning on the device," said Aubry. "We just start work and go about our business; it's all automated."

After work in the field is completed, the data is uploaded, processed, stored, and visualized in ArcGIS Online. Processing involves

The STA logger web app displays the track log and spray zones at Pine Lake.

turning the raw data from the logger into three layers. The first layer is a point layer that shows the location of the device every second it is turned on and contains diagnostic information about the device's status. The second layer is a track log that shows the path the unit traversed while operational, whether spraying or not. The polygon layer, also called the spray zone, represents where herbicide was applied to the ground.

"Before using the STA loggers, we relied on site drawings, written description, and photos to communicate site conditions to clients," Aubry said. "Spatial elements were rarely included in reports, if ever."

Processed layers are appended to a hosted layer in ArcGIS Online and shared with W+P through a web app. W+P uses Esri web mapping tools to export data for reporting and compliance.

"We now embed STA logger spatial data into every report sent to clients," added Aubry. "STA logger data helps us communicate exactly what work we performed on the site inclusive of time, footprints, weed locations, species treated, and spray intensity. We're now able to provide clients site data so comprehensively that it rivals having boots on the ground."

## Conservation through analysis

The STA logger web app also enables W+P to consult STA logger data in the field. When working in an area that has been sprayed recently, a user can quickly determine where the last field supervisor finished and pick up the work from there. Additionally, using the time slider function, the user can observe where weeds were during the last season and focus on those areas.

STA logger data is also uploaded into ArcGIS Pro, where the statistics tools and charts available make it a simple process to gain greater insight into this data and leverage it for future reforestation efforts.

For instance, using the tools available in ArcGIS Pro, W+P estimated that in June and July 2021, W+P operators sprayed 1.78 hectares, or 6.5 percent of the site. This output is useful for any organization involved in the management of weeds and can be repeated at set intervals to track a site's progress over time. Furthermore, it is far more precise and replicable than conventional methods of estimating weed control such as visual appraisal. Users with authoritative data of their site boundaries can use a more accurate polygon and integrate it into automated solutions.

If a particular set of metrics is important for an organization or project, the analysis workflow can be developed into a geoprocessing model, scripted in ArcPy™, or displayed in a dashboard created with ArcGIS Dashboards.

Many statistical outputs can be applied to STA logger data using ArcGIS Pro. Some statistics commonly assessed by STA logger customers include the distance operators travel in a day, the area they spray per day, the ratio of distance traveled to an area sprayed, and the number of trigger presses per unit area (as a proxy for planted tree abundance). Whatever metric is important to weed managers, STA logger data serves as a first step.

The use of geospatially oriented tools such as STA logger and ArcGIS web apps is an important stage in the technological modernization of reforestation and conservation, which in turn has supported practitioners like W+P with tools to create the old-growth forests of the future, one weed at a time.

A version of this story by Harley Schinagl and Mitch Younes originally appeared in the November 2022 issue of *ArcWatch*.

# PART 2

# OUTDOOR RECREATION

OUTDOOR RECREATION ON PUBLIC LANDS PROVIDES A valuable public service to our communities and supports local economies. GIS provides public and land managers with the situational awareness to keep visitors safe. With GIS, environmental organizations can more effectively communicate recreation opportunities, interpret resources, engage with the public through community science initiatives, and ensure maintained and safe recreational assets while responding quickly to changes in status. GIS also helps land managers measure and visualize the economic impact of outdoor recreation and communicate these benefits to all stakeholders.

## Protect natural and cultural resources

GIS supports land managers as they catalog and track natural and cultural resources and ensure their protection before, during, and after recreational development and maintenance.

## Licensing, education, and outreach

GIS helps land managers understand licensing trends, target educational activities, and conduct outreach on priority initiatives.

## Manage assets

With GIS, outdoor recreation managers can prioritize and plan capital projects, track their progress, and ensure staff and visitor safety in our parks and outdoor spaces.

## Assess economic impact

Outdoor recreation is an important economic driver for communities. With GIS, a manager can conduct economic impact analysis, visualize this impact across space and time, and understand who benefits and how to optimize investments in outdoor recreation across their region.

## GIS in action

This section will look at real-life stories about how recreation organizations use GIS to keep parks and visitors safe in the outdoors, more quickly respond to changing situations, provide recreationists with modern tools for exploring public lands, and understand the economic impacts that our public lands provide communities and gain insight on how to best conduct outreach to new recreational enthusiasts.

# LAW ENFORCEMENT TRACKS VEHICLES AND CALLS FOR SERVICE WITH REAL-TIME DATA

## Manassas Park, Virginia

FROM LAW ENFORCEMENT TO PUBLIC WORKS AND PARKS and recreation, the local government of Manassas Park, Virginia, is an independent agency that offers a full range of services to meet the needs of residents and businesses. Recognized as the seventh-safest city in the state, the local government and law enforcement strive to make Manassas Park a secure place for anyone to call home.

Chris Himes, the assistant city manager of Manassas Park, began an initiative to grant citywide departments access to GIS data. The city had to scale back administrative services during the 2008 recession. So the city not only had to transition from its legacy system to help better integrate its siloed data systems, but it also needed the help of technology firm Blue Raster to provide GIS services for the initial migration and ongoing support. Blue Raster, a member of the Esri Partner Network, is helping companies tell their story through interactive mapping technology.

Himes said the immediate need was to get the Manassas Park police division's new fleet an automatic vehicle location (AVL) tracking component device inset for public safety vehicles and attach a computer-aided dispatch (CAD) system for calls-for-service data. The legacy CAD system previously in use did not have an integrated GIS component. Himes began the search for a GIS product to give departments across the city access to GIS data.

Blue Raster tested and deployed ArcGIS Velocity$^{SM}$, giving the city near-real-time data and a map-based view determining the location and status of police vehicles as well as recording and plotting calls for service. The integrated component now gives emergency personnel a comprehensive picture of the city.

## Esri integration

According to Himes, when he began his role with the city, the goal was to develop and implement a comprehensive IT Master Plan to dictate how the city would centralize all services and use GIS to enhance reporting, transparency, and performance management. After examining different options, Himes selected ArcGIS Online because it offered multiple access points for GIS novices as well as city personnel with more familiarity.

The initial part of this project was the migration to ArcGIS Online from the previous software provider. All the stored data was still in a readable format in the third-party web app, and Blue Raster needed to convert it into a cloud-based GIS. Andrew Patterson, a GIS analyst with Blue Raster, said team members began by taking an inventory of all the data that had been delivered to them. Then, they went through the process of publishing the data and organizing it by department in ArcGIS Online. Overall, almost 100 GIS layers were migrated.

Blue Raster then began doing a one-for-one implementation for each city department in need of a geospatial viewer of Manassas Park data. The team had planning meetings with the departments that already had a GIS viewer to get an idea of what their needs were during the transition, such as determining what layers had been available to them and what layers were not in use.

"We then worked through implementing those needs into different ready-made applications using ArcGIS Experience Builder, so that from the [users'] standpoint, they have a named user, and they sign on to the same link every single time. It never changes, no matter how much we update it," said Patterson.

## A move to Velocity

The next phase in this process was shifting the focus to the needs of Manassas Park law enforcement. Patterson explained they wanted to ensure that police had full usability with an AVL and CAD integration and a real-time solution, which is what led the group to Arc-GIS Velocity. ArcGIS Velocity is a cloud-native add-on capability for ArcGIS Online designed to help users process and analyze real-time data feeds.

As the city had already invested in ArcGIS Online, the Blue Raster team believed this Esri solution would be the best option to provide the infrastructure to do near real time, said Patterson. Also, with no dedicated GIS support, Himes liked that it would work in the background with no heavy oversight needed because it was a cloud-native solution and was scalable to fit the needs of the city.

"Because of the city's implementation of ArcGIS Online, if we were to use [another] near-real-time solution instead of Velocity, we would have to have implemented an entirely separate infrastructure to use it," Patterson said. "I think the decision really came down to the cost of it. It made the most sense in every single box."

In addition, the City of Manassas Park already had a Microsoft Azure account, which would help with the Velocity deployment. Patterson explained that because Velocity is based on Azure infrastructure, it meant the city could use a cloud-hosted solution it was more familiar with as opposed to other services. As such, the costs of Velocity could be wrapped up in the city's annual spending for Azure. Blue Raster also had its own environments to test with Azure, so Patterson said everything fell into place.

The Blue Raster team wanted to test Velocity before deploying and putting it into production for the city. Patterson explained that Blue Raster team members used their own cloud environments in Azure, acquire a test license for Velocity, and test all the

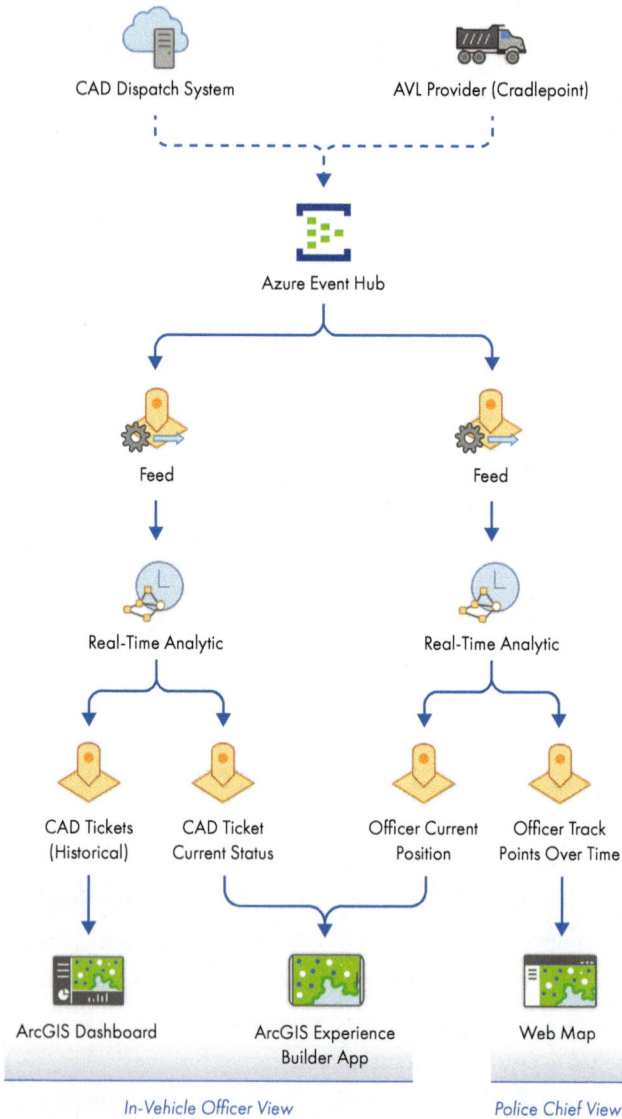

CAD Dispatch System

AVL Provider (Cradlepoint)

Azure Event Hub

Feed

Feed

Real-Time Analytic

Real-Time Analytic

CAD Tickets (Historical)

CAD Ticket Current Status

Officer Current Position

Officer Track Points Over Time

ArcGIS Dashboard

ArcGIS Experience Builder App

Web Map

*In-Vehicle Officer View*

*Police Chief View*

High-level functional architecture integrating ArcGIS and real-time data from public safety information systems.

development—from receiving the AVL to using Azure Event Hubs (a real-time data streaming platform).

## A public safety use case

The primary use for Velocity for public safety in Manassas Park is to have a geospatial viewer of the city map with all linked public facility schematics in PDF files and response mechanisms housed on one page. An integral part of this solution is live layer feeds with locational data from the AVL component that can be toggled on and off and be updated every few seconds. This capability is for officers who are logged on to their Cradlepoint routers, which provide network connectivity in the field and are responsible for the AVL feed, and have their mobile data terminal (MDT) activated.

One live layer feed provides near-real-time data on active police cars. If a car is turned on, the location is plotted on a map and continues up to 30 minutes after the vehicle is turned off. The second feed is for the CAD dispatch system, which is designed for recording and prioritizing incident calls and identifying locations of field personnel, to show when a ticket is logged in to the system. When tickets are logged, the data is processed by Velocity and made available on a map to officers in less than a minute. Active events from the past 24 hours are also displayed.

"We are using Velocity so that whenever we log a CAD ticket, we only log it once. One of the concerns was, with the CAD, when a service ticket is opened, it sends a message once. When it's closed and no longer an active event, it sends a second message. So we wanted to make sure that we aren't getting two points for every one ticket," explained Patterson.

A dashboard was set up to log calls for service, which uses the live feed of CAD data and enables users to search and filter data; for example, viewing only calls for service for parking violations over a

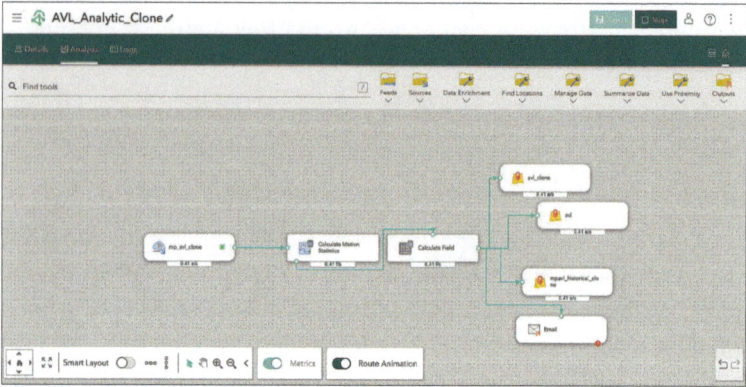

The AVL Analytic feed receives information from in-vehicle routers as the information source.

two-month period. The live feed lets users see active events, calls for service, and the locations of other active units.

## Results with Velocity

The AVL component was rolled out to the police staff and police chief, law enforcement deputies, and E-911 employees, among others. The solution combines a dashboard and mapping interface to easily view data, and Velocity gives the ability to make data consumable in a GIS.

Patterson explained that because the police department data comes from different sources such as the Cradlepoint routers, the data does not easily integrate into a GIS. However, for both data feeds, there are solutions in Azure to receive the data and transform it into a GIS-type format. For example, AVL comes in as a coded message, and Blue Raster parses out the information needed to translate it into a more readable format. Then, it is pushed into Event Hubs, a standard connector from which Velocity can receive data.

The police department also can gather data on when officers

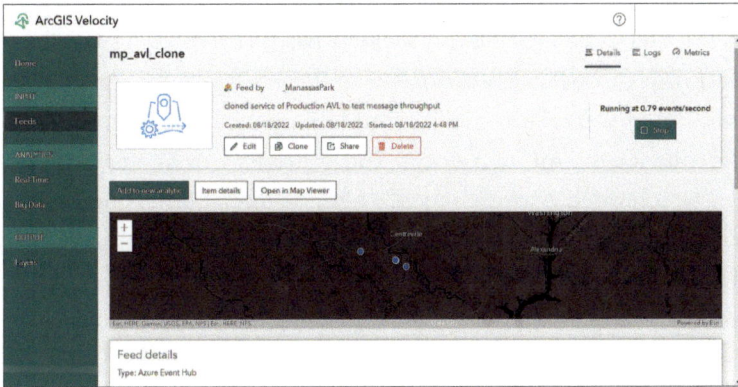

The Velocity feed interface allows a services throughput to be tested without having to toggle between Velocity and a web map in ArcGIS Online. The service can be controlled and checked from the item details page on Velocity.

exceed a certain speed in their vehicle. Velocity gives Blue Raster team members the ability to enrich track data with movement metrics using real-time analytics. Patterson explained that they can trigger events and other information through Velocity, so it receives and enriches data with more details.

"Now with Velocity, I can say…if it's accelerating or decelerating…with real-time analytics to make it useful and not just points dancing around the map," said Patterson.

## Benefits of real-time data

Himes said Velocity has shown the value of GIS-driven solutions in the city and what solutions are possible with Esri technology.

"From what it is right now, it is capable of giving [police staff] exactly what they need from an operational standpoint, and that's perfect for them," said Himes. "But I think if you show that utility just through [continual] enhancement and build-out, it's only going to get more buy-in and use of the service in and of itself."

Patterson added that as a cloud-native solution, Velocity has shown it can provide near-real-time data delivery and enrichment. Moving forward, Himes has several ideas in mind for Manassas Park to enhance the use of Velocity and other Esri technology, including an elevated performance dashboard for managers across the city that allows them to incorporate their data with a mapping view. He would also like to proactively design more customized views for resident services and performance initiatives such as offering the improved ability to report on maintenance issues in the city.

"Having Velocity really made everything else for this possible," Himes said.

A version of this story titled "Law Enforcement in Manassas Park, Virginia, Tracks Vehicles and Calls for Service with Real-Time Data Solution" originally appeared on esri.com.

# DIGITAL TRAIL MAP GETS FIRST RESPONDERS TO TRAIL RESCUES FAST

**New York State Office of Parks Recreation and Historic Preservation, Suffolk County Police, and Cold Spring Harbor Fire Department**

IF SOMEONE FALLS AND HURTS THEMSELVES OR GETS SICK on a 1.14-mile hiking trail in Cold Spring Harbor State Park on Long Island, New York, local police officers, firefighters, and emergency medical technicians (EMTs) come to the rescue.

But for many years, it was difficult to find someone who was reported injured or sick on the hilly, wooded trail in the midst of the 40-acre park in the Town of Huntington, a community of 200,000 people that includes the hamlet of Cold Spring Harbor. People who called 911 to report these kinds of medical emergencies often could not pinpoint their exact location in the woods.

"They just report being on the trail," said James Garside, a Suffolk County police officer who patrols the area. "There was no point of reference. It did leave us with a guessing game."

But thanks to a new system of numbered trail markers erected in the park and a companion map that shows the coordinates of each marker, locating someone who is injured or ill is much easier than in the past. Emergency callers from the trail can now report the number of the trail marker closest to them. And first responders can consult a digital trail map on an ArcGIS Explorer app to obtain the marker's geographic coordinates and additional information that will aid in the rescue.

The information the map provides to first responders shaved about 10 to 15 minutes off the response time to a medical emergency that occurred on the trail, said Garside, who spearheaded the effort to install the trail markers.

The yellow dots show the location of each trail marker.

On October 15, 2017, a 47-year-old man suffered a heart attack on the trail, according to Garside. The man collapsed at marker 108, one of 15 small signs placed less than one-tenth of a mile apart from each other along the rugged trail, which is surrounded by oak, red maple, American beech, and other trees. The heart attack victim's wife called Suffolk County's Enhanced 911 system on her cell phone and reported the trail marker number, posted on a tree that her husband sat slumped under.

Garside and first responders from the Cold Spring Harbor Fire Department sprang into action. Besides being the officer who patrols the area, Garside is trained as an advanced emergency medical technician (AEMT). He consulted the trail marker map, which is available to him on the Town of Huntington's Explorer app and on a data terminal in his patrol car.

The Explorer app displays the trail running through Cold Spring Harbor State Park.

On the map, Garside could see the details important to coordinating a quick response plan: the latitude and longitude for trail marker 108, the best access point to get to that site, and suggestions as to the types of vehicles and apparatuses to use to bring the patient out.

In 2015, Garside had approached the New York State Office of Parks Recreation and Historic Preservation (NYS OPRHP) with the idea of installing the trail markers in Cold Spring Harbor State Park. NYS OPRHP assigned a GIS team to map the trail and gather each sign's coordinates.

The Cold Spring Harbor Fire Department then sent a team to walk the trail and create its own response determinates. These included descriptions of the best places to access the trail to get to each marker location (e.g., the south or north end of the trail or a specific residence "near the barn"), the types of equipment suited to the terrain at that site, and the best extrication point. That information was then added to the trail marker map.

In the case of the heart attack victim, the best access point to reach him was through private property on the 200 block of Harbor Road/New York State Highway 25A. When Garside arrived at the house carrying his medical equipment—a Physio-Control LifePak 12 portable cardiac monitor—the property owner was helpful.

"He was pointing me in the right direction [toward the trail]," Garside said.

Garside reached the heart attack victim's side in five minutes. The typical response time without the accurate trail marker information might have been about 15 minutes.

The Cold Spring Harbor Fire Department used a Mule litter wheel to bring out the heart attack victim—the recommended equipment listed on the Town of Huntington's Explorer app.

But prior to that, Garside obtained an EKG reading from the man and sent the readout to a nearby hospital emergency room. The staff there studied it and called in a cardiac care team, so the physicians and nurses could be in place and ready when the patient arrived at the hospital.

The trail signs, the information available via Explorer, the rescue equipment, and the mobile medical technology—along with the first responders, of course—all helped provide a happy ending to the story.

"It's a blend of old technology, with signs, mixed together with new technology. It worked well," Garside said. "It did save that man's life."

NYS OPRHP's map of the trail that displays markers 101 through 115, along with each sign's coordinates, was posted at a kiosk in Cold Spring Harbor State Park. Visitors can add the trail marker coordinates to their smartphones by scanning a QR Code on the map.

The spatial data from that project, together with the information from the Cold Spring Harbor Fire Department, was shared with the Town of Huntington. The town's GIS manager, Dave Genaway, said he was alerted to the data by Huntington chief fire marshal Terry McNally.

Genaway said the trail marker data was added as a feature layer to the Huntington Fire Preplan app hosted in its ArcGIS Online organization. The Huntington Fire Preplan contains information about local buildings—including roof materials, known hazards, and floor plans—that help firefighters plan their response to fires or other emergencies. That planning and trail marker information also is available to local firefighters and first responders via Explorer for ArcGIS.

When users click on one of the trail markers on the map, the latitude and longitude for that marker appears.

"It also shows which access point they can use to get to that trail marker in [the most] efficient way," Genaway said.

Garside hopes that the trail marker system could be duplicated in other parts of the state in the future. But for now, the police officer is pleased that Cold Spring Harbor State Park will be a little safer for hikers and runners. He said that he came up with the idea for the trail marker system after a man who was walking on the trail with his wife one day in 2015 had a close call.

"It was National Trails Day, the first weekend in June," Garside recalled. "That [incident] was the straw that broke the camel's back."

The man fainted on the trail due to a heart condition. His wife called 911 but had no idea exactly where on the trail they were located.

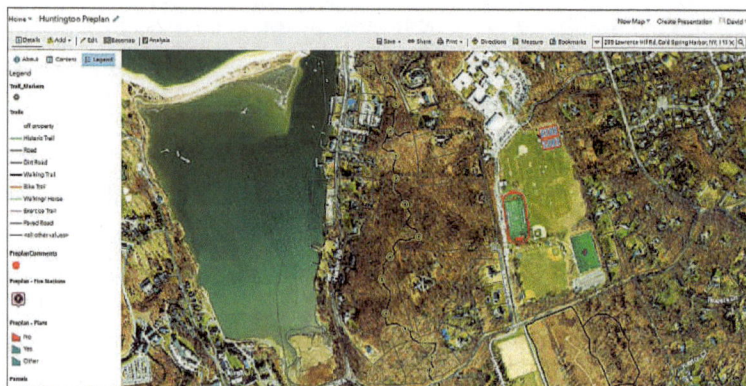

The trail marker data was added as a feature layer to the Huntington Fire Preplan app, which helps firefighters plan emergency responses.

"When you are on the trail, to the right or left you see woods," Garside said, adding that the steep hills and trees make it hard to stay well oriented.

Garside said it took about 10 or 15 minutes to find the ill man on the trail, which is basically a footpath and not accessible by ambulance. Other hikers on the trail tried their best to help but were unable to give first responders a good estimate of how far down the trail the man and his wife were located, according to Garside.

And what about his inspiration for how the trail markers would work? Garside modeled them after the signs on the interstate highways in New York that include exit numbers.

"I just thought, 'There's got to be a better way [that] would take the guesswork out of it,'" Garside said. "These trail markers provide that."

A version of this story originally appeared in the Winter 2019 issue of ArcNews.

# HIKERS GO EXPLORING WITH SMART MAP TRAILS APP

## Greenbelt Land Trust

THE WILDERNESS OF ESSEX COUNTY, MASSACHUSETTS, HAS long inspired explorers and free spirits to wander its paths. Henry David Thoreau, a Massachusetts native known for his reverence of the local woodlands, wrote, "To be admitted to Nature's hearth costs nothing... You have only to push aside the curtain." For residents interested in exploring the vibrant trails of the northeast corner of the state, in pushing aside that curtain, it can be hard to know where to start.

This is the challenge the county's Greenbelt Land Trust set out to meet with its new GreenbeltGo app, detailing and mapping data on hundreds of miles of hiking trails. For 60 years, Greenbelt staff have been working with 34 cities and townships, using GIS technology to map priority places and preserve land throughout the county.

Their work helps promote sustainable ecosystems and maintain equitable access. "The core goal was making our properties as accessible and welcoming to as many people as possible," said Abby Hardy-Moss, director of conservation technology and planning at Greenbelt. "We wanted to help people be comfortable out in the woods and at all the Greenbelt properties."

Hardy-Moss and others within the organization knew visitors were ready for a mobile tool for finding and using trails. "We already had interactive web maps for all our properties online, as well as a countywide one," Hardy-Moss said. "The maps are an extremely high-traffic part of our website."

The app advances the online maps by offering trail selection, live navigation, up-to-date advisories on changing trail conditions, and information about the features that make each trail special.

## Deciding to make an app

Visitors seeking to experience the natural and historical wonders of Massachusetts can find such features as Chadwick Pond on the Bailey Reservation, where Algonquians fished long before European contact, and Clamhouse Landing on the Allyn Cox Reservation, now a river headland, which began as a shell heap started by Pawtucket clam diggers over 2,500 years ago.

Before Greenbelt released its map-based app, people who wanted to explore the lands under Greenbelt's conservation had to consult static trail maps or an interactive map on the land trust's website. The data was kept as up-to-date as possible, but information on trail conditions was still limited, and these maps couldn't effectively help hikers orient themselves while they were on the trails.

"We talked to a number of partner land trusts and similar organizations who had created trail apps," Hardy-Moss said. "We spoke with people from Allegheny County GIS, who made an incredibly generous offer to share a lot of their existing code with us, and now if we have updates, they can use that and vice versa."

Working with their colleagues at Allegheny, Greenbelt developers moved quickly. By June 2021, they were already working on GreenbeltGo, with a goal of launching the new solution by early spring 2022. "We could see a clear pattern of visitors to our maps based on the season, so we wanted to try to release the app by spring, because that's when people really start getting out onto our properties the most," Hardy-Moss said.

## Taking to the trails

To ensure the foundational information was as current as possible, the Greenbelt team went into the field to collect trail data using GIS equipment. The endeavor required them to hike 120 miles of trails in about six months. A great deal of this was accomplished by

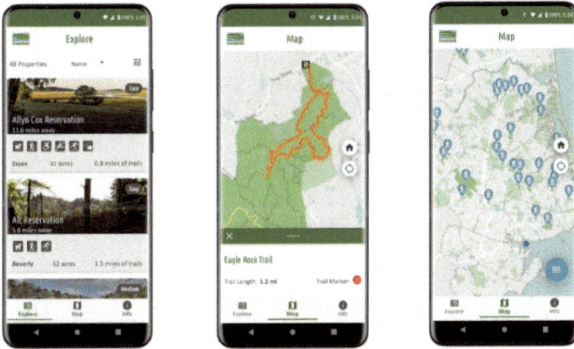

Users can download the GreenbeltGo app to their phone to navigate open spaces.

Rebecca Smalley, a senior at Salem State University and Greenbelt's GIS intern during this project. Smalley, who is pursuing a degree in GIS and cartography, was also responsible for editing and organizing the resultant data.

"The most rewarding part about this whole experience was being able to physically see the trail data I collected and edited in the app and knowing that so many people are already enjoying it," Smalley said. "I'm confident that the app will encourage people to get outside and understand the importance of land conservation."

GreenbeltGo was released on February 14, 2022, covering 49 protected properties as well as state-owned properties such as Bradley Palmer and Maudslay State Parks, with plans in place to expand to additional areas in the future. People can plan their visits by type of activity, proximity to their community, and degree of hiking experience. Via an interactive map, users can explore points of interest and elevation contours, access up-to-date information on trail conditions, and track their locations on the trails live as they go. They can download offline maps if desired and receive notifications on upcoming outreach and educational events organized by Greenbelt.

By placing a new level of detail on natural and cultural resources at hikers' fingertips, GreenbeltGo acts as an advanced guide for local exploration. "So far, the feedback has been really positive," Hardy-Moss said, who notes a lot of interest and engagement from the community.

## Enabling two-way communication

The app also provides a way for visitors to participate in the stewardship of the Greenbelt lands, using the power of precise location and technology to communicate clearly.

"Before, people would report a tree down or another issue, and our people would have trouble finding the spot they were talking about," said Dave Heacock, Greenbelt's geographic and technical support specialist. "Now, people can report trail issues using their actual location and pictures of the problems—data for us to act on."

This app feature ensures the most accurate and immediate information possible on trail issues while giving visitors a new way to truly be invested in the welfare of Massachusetts's natural spaces.

"My biggest takeaway, and the thing I'm most excited about, is just people being more confident visiting our properties," Heacock said. "People are potentially using trails and properties more often than they were before, because the app is easier to use and more trustworthy and up-to-date than a web map or handheld map."

A version of this story by Sunny Fleming titled "Massachusetts Hikers Go Exploring with Smart Map Trails App" originally appeared on the *Esri Blog* on May 26, 2022.

# COVID-19: STATES APPLY LOCATION INTELLIGENCE TO MANAGE PARKS

**Tennessee State Parks**

WHEN CORONAVIRUS DISEASE 2019 (COVID-19) STAY-AT-home orders first began to pop up throughout Tennessee communities, the state's parks saw a sudden increase in visitors.

Alarmed by the huge number of local and out-of-state guests visiting during that first week, park managers grew concerned about potential health risks for visitors and staff. Tennessee governor Bill Lee quickly responded by closing the state's 56 parks. At the time, officials were uncertain of when and how to safely open them again.

Staff at Tennessee State Parks needed a way to help the governor's office monitor the COVID-19 crisis, make policy, and create guidelines. Moreover, they needed a system that would help state park managers plan reopening strategies. They also wanted to keep the public informed about which parks were open and what restrictions were in place.

Since location plays a crucial role in COVID-19 safety and prevention, state leaders decided to use a GIS to monitor and manage the crisis. For years, Tennessee State Parks had been using the technology, and a team of professionals had kept up with location intelligence advancements. They had already built numerous GIS applications for managing park and business operations, so it was not a stretch for the department to create a GIS strategy specifically designed to handle the pandemic.

## Managing information essentials

Former Parks and Conservation GIS Lead Rachel Schultz and a team at the Tennessee Department of Environment and Conservation created a location strategy that would help leaders launch safe initiatives. First on the to-do list was to determine each park's risk factors and levels. Next was to create a dashboard that would give the governor's office actionable information. This information would be updated throughout the day as needed.

"Previous park closures were due to budgetary and economic concerns," Schultz said. "These current and partial closures are caused entirely by public health concerns. It's a new arena for us, as I believe it is for many organizations."

Simultaneously, the GIS team expanded park staff's technology capacity, developing apps and analysis to provide decision-makers with the location intelligence they needed. The team used a variety of data to support these tools, including existing information on parks, COVID-19 tracking data from the state health department, and modeling output data from the Institute for Health Metrics Evaluation (IHME). But the most valuable data came from the parks themselves, especially real-time information captured by park managers and rangers using ArcGIS mobile apps for data collection.

## Analyzing risk to reopen parks

Risk analysis plays a large role in understanding and managing response for state parks. To determine COVID-19 risk levels for each park, the GIS team created a web survey app to collect information from park managers. It asked them to rate their concern levels about continuing regular operations during the outbreak as low, moderate, or high. The survey also asked what specific situations were driving those concerns.

The team created an operations dashboard that showed the spatial distribution of parks with high concerns. It also used geographically weighted regression tools to find patterns that explained why some parks were more or less concerned with risk than others.

Findings showed the highest concerns came from parks with rugged or steep terrains in which park rangers would be exposed to COVID-19 during rescues. Also, parks that had sufficient stocks of personal protective equipment (PPE) and cleaning supplies were more likely to give a low or moderate concern rating.

## Deciding when to reopen parks

Staff used a GIS park sustainability dashboard to brief Governor Lee on the status of Tennessee State Parks. This allowed him to work on epidemic response including the difficult decision to temporarily close state parks.

During the closure, the team completed two more GIS projects. The first one tracked PPE and cleaning supply inventory at parks across the state. It helped management determine how best to distribute equipment and supplies among state parks. A dashboard measured status of the goal of getting each park fully stocked with everything they needed to feel secure upon reopening.

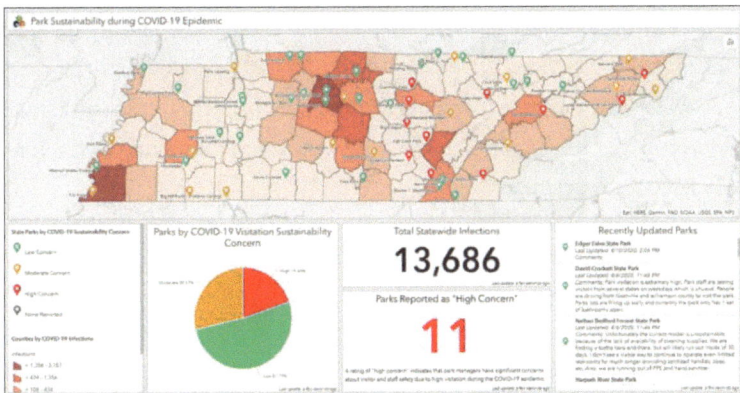

The Tennessee State Parks COVID-19 dashboard gives decision-makers a means to view summaries for all parks related to sustainability during the pandemic.

Within the parks, staff posted signage about closed trails, social distancing rules, and face mask requirements.

## Informing the public

Tennessee State Parks wanted to supply people with information that would be easier to read than a simple list. The GIS team lead worked closely with the park's marketing team to embed an ArcGIS web app at TNStateParks.com. They built a publicly accessible, interactive map that interfaced with the GIS database.

Whenever a park's status changes, the map automatically updates and shows the most current information. If a park experiences an unsustainably high amount of visitation over a weekend, the park can close or restrict admission and immediately show it on the public map.

A tenet of social distancing is to stay close to home. Although Tennessee encourages people to come out and enjoy state parks, it asks visitors to limit travel as much as possible. Fortunately, all Tennesseans can find a state park within a one-hour drive. The status map will quickly and clearly show the nearest park and its current status and restrictions.

"We have never experienced anything like this," Schultz said. "GIS helps us to respond to these new and ever-evolving situations quickly and efficiently."

A version of this story by Mike Bialousz originally appeared on the *Esri Blog* on June 17, 2020.

# STRIKING A BALANCE BETWEEN CONSERVATION AND GROWTH

## Thrive Regional Partnership

B EFORE THE PANDEMIC NORMALIZED WORKING FROM home, Chattanooga, Tennessee, with the nation's fastest broadband, was lauded as the best city for remote workers. While the area's mix of ecology and technology is attracting new residents, the surge is triggering development pressures that may disrupt the balance.

Much of Chattanooga's appeal is its proximity to outdoor recreation, enabling what *City Journal* called a "rock-climbing-before-work, kayaking-before-dinner lifestyle." The city's locale—tucked between the Cumberland Plateau and the Blue Ridge Mountains—means these activities occur alongside some of the nation's most ecologically rich and fragile areas. Only California has more threatened biodiversity.

"The geography in the region is really unique, and the variation in landscape and topography drives the biodiversity," said Charlie Mix, director of GIS technology at the University of Tennessee at Chattanooga.

Maintaining a healthy balance between conservation and growth is an ingrained local ethic in Chattanooga, perhaps best exemplified by the large auto assembly plant Volkswagen announced in 2008 and opened in 2011. It occupies one corner of a 6,000-acre plot, with nearly half of the remaining land dedicated to a public nature park.

## Learning to thrive

Volkswagen's arrival intensified the conversation around economic development and protection of natural assets, spurring the creation of Thrive Regional Partnership (Thrive), a nonprofit organization that inspires responsible growth in the tri-state region.

Thrive unites various partners such as government officials, private-sector leaders, and academics across northeast Alabama, northwest Georgia, and southeast Tennessee to ensure that as the region grows in population and industry, community and natural character is preserved.

"In an initial regional planning initiative that began in 2008, we learned that the tri-state was projected to grow by half a million people by 2055," said Rhett Bentley, strategic communications director for Thrive. "People move to this area not only for economic opportunity, but for access to beautiful natural spaces and scenery. Leaders from across the region created Thrive to balance development with the natural treasures and nature-based lifestyle of the region."

Rather than consider just the city of Chattanooga or its extended metro area, Thrive employs what could be called a "micro-mega-regional" approach. Its mission covers a 16-county area (including seven in Georgia and Alabama) that Thrive calls the Cradle of Southern Appalachia.

The Natural Treasures Alliance, one of Thrive's ongoing coalitions, advocates for long-term policies that address landscape conservation and preservation. Around 14 percent of the Cradle's land is currently protected. The project aims to double that number by 2055, aligned with the global 30 by 30 movement, which aims to protect and conserve 30 percent of the world's land and sea habitat by 2030. For the Cradle, that amounts to one million acres of new land set aside for conservation. In a region with a population of a million people, where outdoor recreation is a staple of local economies, this is no easy task.

"Outdoor recreation brings in hundreds of millions of dollars annually through tourism and consumer spending according to the

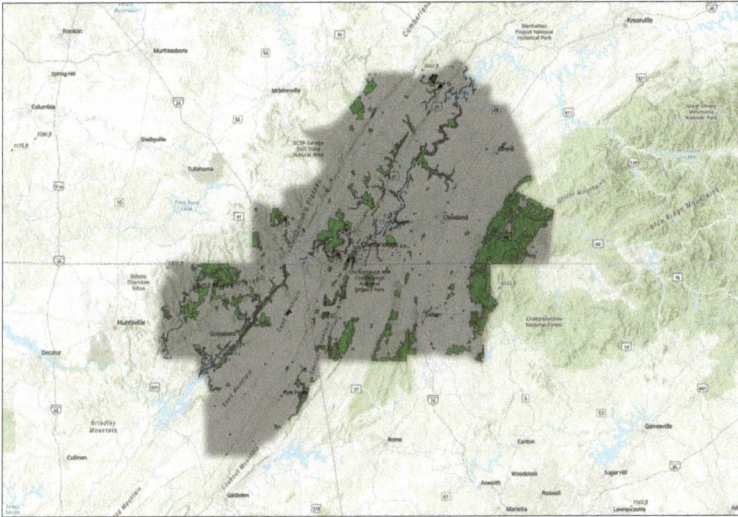

This map illustrates the protected areas (in green) in the Cradle of Southern Appalachia as of 2021.

Outdoor Industry Association," Mix said "It's arguably one of the most important parts of the regional economy."

The aquatic biodiversity alone makes the Cradle ecologically unique. A major goal of the Alliance is to get 50 percent of the streams currently considered imperiled removed from the federal government's list of endangered waterways.

"Freshwater mussels are considered one of the most imperiled fauna groups on the continent, and Tennessee is home to over 100 species," said Matt Reed, the natural treasures program director for Thrive. "Alabama and Georgia have more species of fish than any other state."

Though the region is rich in biodiversity, much of its land is unprotected, according to Joel Houser, southeast field coordinator for the Open Space Institute and a member of Thrive's Board of Trustees.

"We recognized that we needed to put forth a bold vision that spatially identified areas important for conservation," Houser said.

## Natural treasure hunt

To carry out its mission, the Natural Treasures Alliance uses a landscape conservation model developed collaboratively with scientists, conservation leaders, and stakeholders. The model integrates various datasets, placing higher weight on land best suited for conservation. Model inputs include habitats where biodiversity thrives, current protected areas, and places with high resiliency to climate change. It also considers connectivity, preserving areas that form wildlife corridors that foster genetic resilience in wildlife.

For instance, one of the corridors, which the Alliance calls the Appalachian Connector, links the Cumberland Plateau and the Blue Ridge Mountains—the same geographical features between which Chattanooga has developed. The Alliance's model prioritizes the beltway that allows wildlife to bypass urban areas unimpeded.

The resultant map of the Cradle displays areas of highest conservation value to serve as a guide to focus efforts and prioritize land purchases and conservation projects.

Besides helping to formulate plans of action, the map is also a way to measure progress. As lands come under protection, the map reflects the changes. Students at the university's Interdisciplinary Geospatial Technology Lab, acting as data stewards, use the map to add to the national database of protected lands maintained by the US Geological Survey.

## The power of simple models

"What we put into this model was intentionally simplistic, but it follows years of data collection and trying to understand the data available for the region," Houser said. "Because of the complexity of the data, we recognized that we needed something simple for the model."

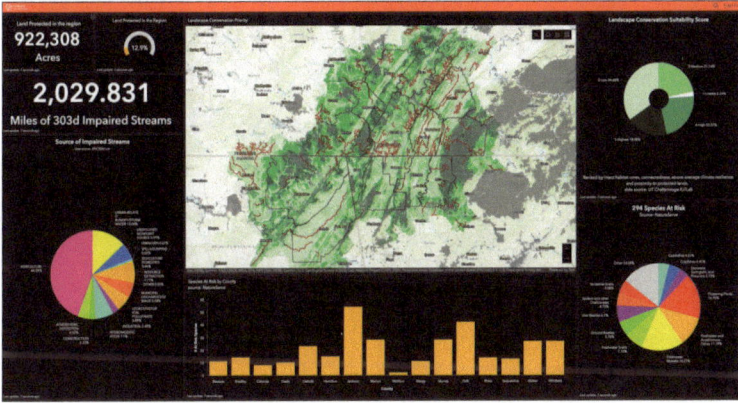

The Cradle of Appalachia Conservation Dashboard provides a continuously updated progress report on areas conserved by the Thrive partnership.

As a visualization technology, GIS gives power to diverse stakeholders to collaborate using one cloud-based map, and the public-facing component helps the Alliance promote its work. The Cradle of Appalachia Conservation Dashboard provides a continuously updated progress report on the project.

The map serves as a visually intuitive way to explain what progress means for the region. "When we share what we're doing with city planners and regional transportation organizations, the map is a key component," Reed said. "Otherwise, we're just sharing data and charts."

Civic leaders in Chattanooga have applied the model to city parks, seeking out biodiversity in an urban environment. The Alliance's model and geographic approach can serve as a guide for screening conservation projects and predicting their efficacy at protecting biodiversity.

Houser sees the work of Thrive and the Alliance resonating throughout Appalachia, especially as climate change and biodiversity loss force communities to adapt new strategies for environmental protection and human survival.

"We're thinking about the importance of the south-to-north corridor that runs from Alabama to Maine," he said. "We're really at the base of it, the cradle. And what we do here has outsized impacts moving north."

A version of this story by David Gadsden and Sunny Fleming originally titled "Striking a Balance Between Conservation and Growth — Maps Show How" appeared on the *Esri Blog* on July 12, 2022.

# BOOSTING PARTICIPATION IN OUTDOOR RECREATION

## South Dakota Game, Fish, and Parks

SOUTH DAKOTA GAME, FISH, AND PARKS (GFP) FOCUSES on connecting people of all ages to the outdoors. This mission aligns with South Dakota governor Kristi Noem's Second Century Initiative, created to enhance wildlife habitat while getting families involved in outdoor recreation.

In lockstep with the governor's new initiative, GFP continually looks for strategies to increase engagement and participation by making it easier for people to access state hunting, fishing, and park resources.

Facing a decline in participation throughout South Dakota, the agency's recruitment, retention, and reactivation (R3) strategies have been elevated to a department-wide priority. The agency has long been a user of GIS technology for data management, field mobility, and analysis for its aquatics, wildlife, and parks.

When Esri released the new ArcGIS Solutions for Recreation License Outreach, agency staff were interested in how it could help them make sense of licensing data through visualization of trends and patterns. The solution laid the groundwork for new campaigns, which would help GFP better engage highly targeted audiences and generate new interest in outdoor program offerings.

## The challenge

Decreasing participation in outdoor activities presents a serious challenge for natural resources and recreational organizations everywhere, and GFP is no different. Considering that hunting and fishing licenses serve as primary sources of revenue, the declining numbers directly impact the agency's ability to implement and maintain important outdoor recreation and conservation programs. This downward trend also damages and deteriorates the health of local economies, job growth, wildlife habitat restoration, preservation of cherished species, sustained history and traditions, and more.

The agency's core offering is unlimited licenses, which has seen a steady decline year over year. Fishing has dropped by 5,000 participants per year from 2016 to 2018 and by 10,000 participants in 2019 alone. Small-game hunting has decreased by nearly 16,000 participants from 2009 to 2018.

To revitalize critical numbers, agency staff needed to gain a better understanding of the downward trends along with potential countermeasures. They lacked an adequate process and workflow to analyze data from their licensing system.

"We didn't have any baselines or data collection or analysis systems to provide a comprehensive, statewide picture of the situation," said Taniya Bethke, former R3 coordinator and education division staff specialist. "We were using different evaluation and data reporting systems for the outdoor program numbers, which meant there

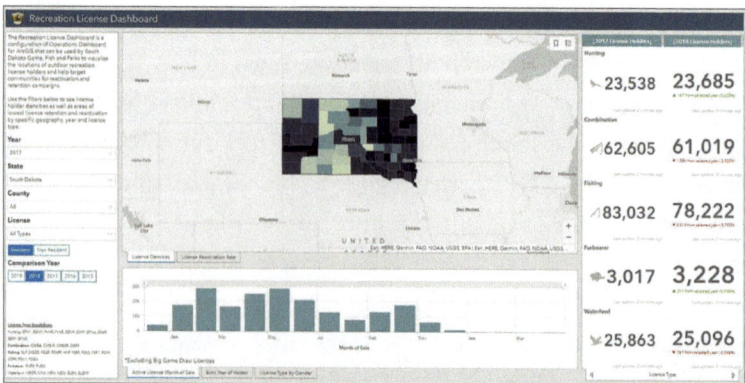

ArcGIS Solutions for Recreation License Outreach enables South Dakota GFP to create dashboards to visualize the number of outdoor recreation license sales by specific license type over time.

was no common view, shared understanding, or real collaboration from department to department."

## The solution

GFP staff turned to ArcGIS Solutions for Recreation License Outreach to help aggregate, manage, and analyze foundational data from their license permitting systems. This new approach enabled users to generate critical statistics and visualize trends spatially to make sense of the vast amount of licensing data.

These new capabilities enabled agencywide teams to glean insights and actionable intelligence from the numbers they were seeing, empowering them to create and manage highly effective R3 outreach campaigns.

"We needed to enhance our offerings for outdoor families by driving the elevation of R3 in our state to a department-wide concern," said Bethke. With ArcGIS, the agency's emphasis has shifted beyond just education and outreach. "It now includes securing new licensing opportunities; performing evaluations of our regulations; providing additional access for hunters and anglers in the field; and ensuring that resources, such as loaner equipment, are made available to them in the right location."

Kyle Kaskie, a former GIS program specialist, was the first to discover how ArcGIS Solutions for Recreation License Outreach could provide the tools they needed to overcome their challenges. He presented the plan to the agency's leadership team, outlining their newfound understanding of how the licensing numbers directly relate to specific demographic and geographic information.

"The agency's management team was very pleased with what we were able to bring to the table and expressed interest in moving forward with the solution," said Kaskie. "After the initial presentation and adoption, we have seen a great deal of insightful questions and input relayed to our development team. That has helped the R3 application continue to grow and mature, bringing even more value to our foundational data."

## The results

For the first time ever, GFP staff can visualize license data on smart maps and dashboards in real time, discovering answers to questions they didn't know to ask before.

This was a direct result of the location intelligence provided by ArcGIS Solutions for Recreation License Outreach. New purchases or renewals, illuminated as hot spots on smart maps, can be sorted and viewed by location, demographic, or specific outdoor activities through Esri's ArcGIS GeoEnrichment℠ Service. Agency staff and partners can now provide specific recreational programs, such as hunting or fishing learning courses, by pinpointing potential participants in localized target markets.

"Having a geographic component to our program licensing trends visualized on one easily accessible map or dashboard is a game changer," said Bethke. "We can now look at a smart map and, in one glance, identify which county my team needs to focus on in 2020 by license type."

The new solution gives Bethke and her team the ability to be intentional with data collection and strategy.

"Gaining new insights from the data—not just viewing the numbers but seeing the causal patterns and trends—is a critical advantage we are using to strengthen our agency's marketing and promotional campaigns, driving engagement, awareness, and, ultimately, greater participation in outdoor recreational activities," she said.

In addition to helping to guide more effective work for the agency, ArcGIS Solutions for Recreation License Outreach also drives new efficiencies through smarter workflows and streamlined

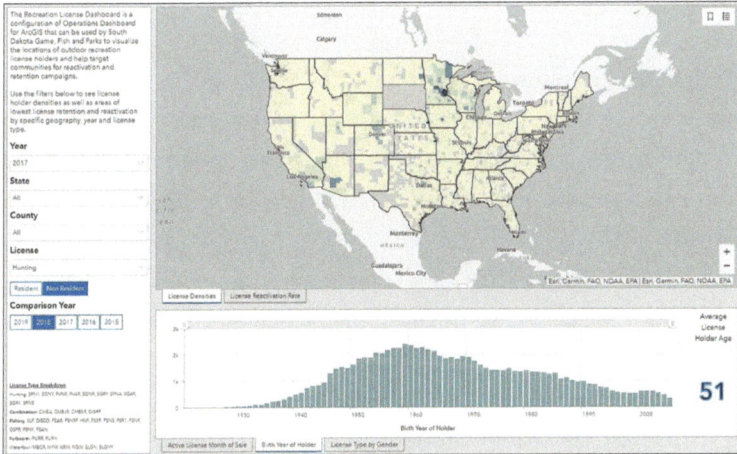

Using ArcGIS Solutions for Recreation License Outreach, the agency is now able to see license holder densities as well as areas of lowest license retention and reactivation by specific geography, year, and license type.

interagency processes. For example, when personnel discover a new opportunity for urban fishing development or a supported hunt in a specific location—illuminated on their smart maps—they can use that data-driven intelligence for more effective outreach efforts.

Location intelligence also enables campaign managers to follow up with well-informed messaging to targeted audience segments in a precise area—informing them that specific program offerings of interest are available. They can create tailored educational messaging to help drive people to a nearby location where they can learn a new skill they have indicated interest in.

"With the R3 solution, our agency is able to determine what the most effective outreach campaign will be," said Bethke. "That wouldn't be possible without the GIS-powered tools to identify potential program participants in a specific community, based on geospatial and demographic data, visualized on smart maps and apps."

## The road ahead

With ArcGIS Solutions for Recreation License Outreach displaying licensing data in a geospatial context, the agency will be able to perform even deeper geospatial analysis—on more detailed and localized audience segments—whether they are existing or potential participants.

This new approach will also help address previously identified needs of a specific type of outdoor participant, based on the participant's stage in the adoption process of new outdoor recreation activities.

Stronger geospatial analysis produces actionable location intelligence. This new level of data-driven intelligence informs effective communication strategies and channels for outreach campaigns, which will resonate with targeted audiences, resulting in increased engagement and participation. Newfound successes will enable the agency to continue to grow, evolve, and implement outdoor programs.

"We want users to be better informed with access to licensing data that we can lay out interactively—filtered by activity, location, or year—to pinpoint exactly where new opportunities are located," said Ross Scott, division staff specialist and GIS coordinator. "We will be able to provide that data instantly to specific users, who can bring that information to commission or executive team meetings or to brief the governor, for example."

These new capabilities are sure to lead to a bright future for outdoor recreation across the state of South Dakota.

"This is something that has never been done before by the GFP," said Scott. "No other state agencies are doing anything like this. We are setting the standard here for licensing. The sky's the limit when it comes to availability and use."

A version of this story titled "South Dakota Game, Fish, and Parks Boosts Participation in Outdoor Recreation" originally appeared on esri.com in 2020.

# THE ECONOMIC VALUE OF PARKS

## The Trust for Public Land

"**W**HAT IS COMMON TO THE GREATEST NUMBER GETS THE least amount of care," Aristotle stated in 350 BCE. In 1968, ecologist Garrett Hardin expanded on this idea, coining it "the tragedy of the commons." Hardin argued that, when it comes to using earth's natural resources, individuals will always act in their own best interest.

The Trust for Public Land (TPL) promotes the benefits of the commons, combating the defeatist sentiment of Hardin's statement that all shared spaces are overused. TPL passionately works to catalyze communities to be healthier, more livable, and more connected through parks and public lands.

While the tragedy of the commons may help explain some negative results of modern life—unsustainable development, air pollution, carbon emissions, depleted water supply—TPL uses data to show how parks and public lands not only mitigate these harmful outcomes but also create shared, positive outcomes. TPL uses conservation economics in a recent report to prove that public parks—an often underfunded public good—provide a myriad of benefits that have monetary value.

To publish *The Economic Benefits of Parks in New York City*, a team of economists, specialists, and research partners used GIS technology to measure the fiscal impacts of city, state, and federal parks within New York City. The report explained how parks lower health-care costs for people who exercise there, provide natural air and water filtration, increase property values, and promote tourism while serving as a place for people to connect with nature.

The report—the first of its kind to study the city's integrated park system—reveals that policy makers have extraordinary financial

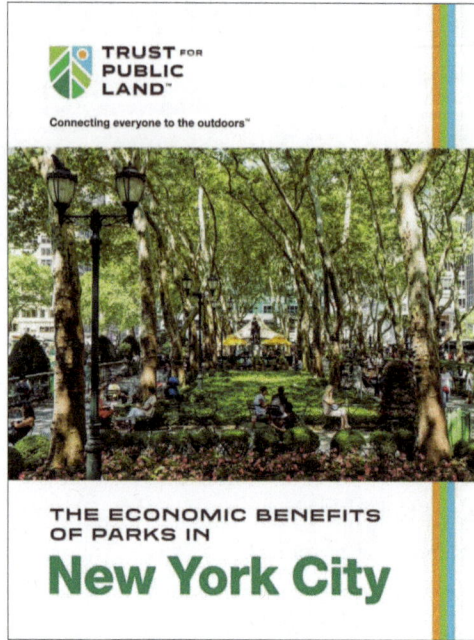

TPL's report reveals that policy makers have extraordinary financial incentive to increase funding for public parks.

incentive to increase funding for the creation, protection, and maintenance of public parks, because parks provide an economic engine for the city.

## Choosing the right tool

TPL is a national nonprofit that works with communities to create parks and protect land. For more than 10 years, TPL's ParkScore index has used information about access, amenities, investment, acreage, equity, and a GIS model to rank how well the 100 largest US cities are meeting the need for parks. Currently, 99 percent of New York City's nearly 8.5 million residents live within a 10-minute walk of a park.

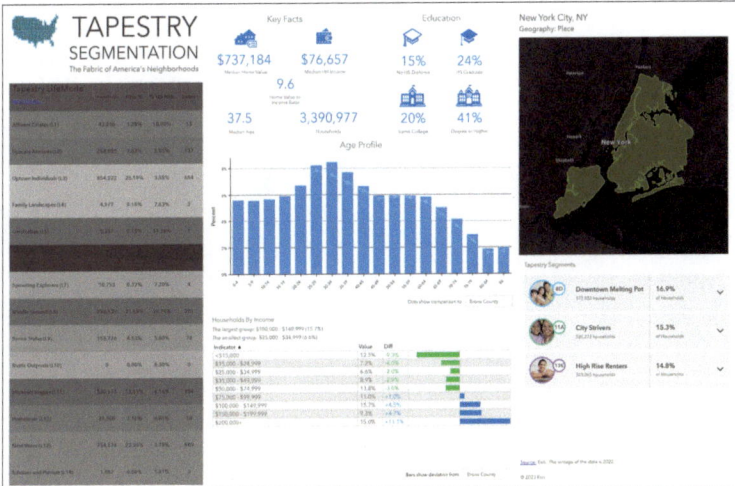

TPL used Esri Tapestry™ Segmentation to analyze New York City neighborhoods as demographic and socioeconomic segments for market and park customer analysis.

ParkScore measures accessibility, but TPL needed a way to measure the intangible benefits that come from park proximity. An understanding that the economy, social systems, and the natural environment are intertwined prompted team members to ask: How can the public health advantages be quantified? How can the benefits of the natural environment be valuated?

There's a growing global interest to put a value on ecosystem services—the many life-sustaining benefits we receive from nature—but nobody had made these calculations across New York City's urban park system.

"We wanted to make sure people understand that, in addition to having a nice place to play and a break from the urban experience that a park offers, there are other indirect benefits in relation to air quality, stormwater management, and wildlife habitat. We wanted to prove that we can make the most of limited funding and

Business Analyst reveals facts about New Yorkers, including median age in each neighborhood and how they spend their money.

meet multiple needs with parks, because they're a big part of what makes a community healthy," said Mitchel Hannon, TPL GIS program director.

The project was no small task. New York City maintains more than 48,000 acres of parkland across the city's five boroughs, making parks a valuable public asset and a critical part of the city's infrastructure. New York State also maintains important recreational assets such as Shirley Chisholm State Park; the National Park Service oversees iconic places such as Liberty Island and Grant's Tomb; and there are many multijurisdictional parks like Governors Island and Hudson River Park. "Valuing ecosystem services and outdoor recreation is a challenge because the information can be so dispersed," said Jennifer Clinton, TPL senior parks and conservation economist.

To analyze and appraise such a vast resource, Clinton, Hannon, and their partners looked to geospatial tools, such as ArcGIS Business Analyst™, which provided location-driven market insights. Business Analyst contains data on consumer behavior, leisure, and business activities in a geographic context, which helped the team estimate how recreation spending contributes to the local economy.

"On a local level, Business Analyst is an obvious tool because of its ability to analyze market potential," Clinton said. "It gave us a way to figure out how residents were spending their money, how often they're participating in certain activities, and the estimates of spending in different categories. We were able to articulate data in a way that allowed us to speak to local policy makers in a way we hadn't been able to before."

## Adding up billions in benefits

The three sections of TPL's report—human health, nature's services, and economic impact—outline billions of dollars in benefits and savings that New York City parks give residents, businesses, and visitors each year. Some of the most impressive figures include the $9.1 billion in recreational value, $2.43 billion in avoided stormwater treatment costs, and $17.9 billion in tourism spending that the New York City park system generates.

"Even though we believe these values are present in our work, we're always a little blown away by the valuations we find in some of these reports. Billions of dollars are reflected in the work of conservation. It's a lot more than people would imagine, in terms of environmental services and economic impact," Hannon said.

For policy makers, the report is timely. The passage of the Great American Outdoors Act in 2020 and the Infrastructure Investment and Jobs Act in 2021 and the introduction of the Community Parks Revitalization Act in 2021 show that parks are a national priority. Communities are well-positioned to use federal funding to reduce the effects of climate change, improve collective health, and boost their local economies by maintaining and creating parks.

The TPL report provides a framework that TPL will replicate elsewhere to quantify the vast potential benefits of parks as well as inspire imagination and motivation to continue and expand park funding.

TPL park projects in New York City.

Because the report's analysis was limited to the years before the COVID-19 pandemic, it serves as a conservative baseline. The TPL team knows that there was a substantial rise in park use in 2020, because they were among the few places people could escape to. Clinton is hopeful that the increase is sustained and continues to generate value and engagement with nature.

## Establishing a framework for the future of parks

In the months since its publication, *The Economic Benefits of Parks in New York City* has influenced policy makers' short- and long-term funding decisions. "Our most immediate need was securing short-term funding for cleaning and maintaining parks that saw record use during the pandemic. But to meet long-term goals of park equity and climate resiliency, we also need to build and maintain new parks. The big structural challenge is finding strong and dependable funding for parks," said Carter Strickland, TPL's vice president for the mid-Atlantic region and New York State director.

New York City mayor Eric Adams pledged to devote 1 percent of the city's budget to funding public parks and broke ground on 104 previously paused park projects in March 2022. In addition to connecting residents to nature, parks appeal to tourists and give business leaders a reason to establish an office or storefront nearby.

"There's a dual benefit to parks—they help bolster the economy, and they're essential for residents' quality of life," Strickland said.

Strickland, Hannon, and Clinton are optimistic about the role of parks in economic recovery, especially in the wake of the pandemic. Appraising the often unseen but important benefits of these spaces provided a critical foundation for their creation, protection, and maintenance. Hannon said this analysis couldn't have been done without GIS.

"It's so powerful to look at a map and see where you need to work," he said. "It's not just theoretical. You're there—right there."

A version of this story by Sunny Fleming and David LaShell titled "The Economic Value of Parks: NYC" originally appeared on the *Esri Blog* on January 19, 2023.

# PART 3

# ENVIRONMENTAL REGULATION

ENVIRONMENTAL ORGANIZATIONS ARE TASKED WITH keeping our environment and natural resources safe. They use GIS to assess project impacts, inform and implement environmental policy, facilitate community engagement, monitor assets in real time, and provide transparency for better outcomes for everyone. The geographic approach allows organizations to track the health of the environment, assess disproportionate impacts on the community, identify the source of pollutants, and prevent environmental hazards from becoming a disaster.

## Ensure compliance

Environmental organizations can ensure compliance through streamlining business processes. With GIS, staff can conduct paperless inspections offline, assets can be monitored remotely in real time, the status of processes can be monitored, and tasks can be automated.

## Enable environmental justice

With GIS, environmental organizations can analyze when, where, and how environmental racism occurs. Location allows this information to be combined with demographic data to understand disproportionate impacts on communities, achieve environmental equity, and engage the public on these important topics.

## Communicate and implement policy

Environmental organizations can use GIS to communicate dynamic environmental policies and permitting requirements no matter where a project may be located. Organizations can use and share tools to assist project planners, consultants, and others to collect and analyze data, apply for permits, and conduct follow-up inspections in the field.

## Monitor environmental assets

Environmental organizations use real-time apps and dashboards to understand data related to environmental quality, such as air and water purity. Using GIS to explore complex datasets allows environmental organizations to easily uncover significant patterns, trends, correlations, and relationships.

## Measure and report progress

Organizations can use ArcGIS StoryMaps and mapping apps to easily communicate with the public, executives, and other organizations the potential benefits of projects as well as their impacts and outcomes on the environment. These location-based tools enable users to quickly measure and report on outcomes and progress.

## GIS in action

This section will look at real-life stories about how environmental management organizations, as well as government entities such as the White House and the National Oceanic and Atmospheric Administration (NOAA), use GIS to help communicate authoritative environmental data to communities for resilience planning, offer transparency around environmental issues such as water quality or how and where funding is spent and distributed, streamline key processes such as environmental reviews, and assess best sites for renewable energy projects to facilitate building a more sustainable future.

# PORTAL HELPS COMMUNITIES ASSESS EXPOSURE TO CLIMATE HAZARDS

## The White House and the National Oceanic and Atmospheric Administration

**F**INDING COMMON GROUND ON OUR MOST COMPLEX issues—including how to collectively act on our changing climate to safeguard our common home—starts with a foundation of shared knowledge, expertise, and information.

As the US advances on its climate action goals with the passage of a massive legislative package, the White House is debuting a new tool—developed with NOAA, the Department of the Interior, and the US Global Change Research Program—that offers scientific data to help us better understand, at a local level, what's happening with our changing climate.

Esri collaborated with the Biden administration on the new Climate Mapping for Resilience and Adaptation (CMRA) Portal to help cities, counties, states, and tribes make better decisions about where and how they need to act. Central to the portal, colloquially being called "camera" by its creators, is the CMRA Assessment Tool; with it, you can explore current and projected climate conditions where you live and work.

With the historic investment to overhaul infrastructure in the United States under way, the country needs to ensure that roads, bridges, railroads, power grids, water systems, transit routes, airports, and ports achieve climate resilience for all people in all communities. This challenge requires a broad understanding before undertaking the tasks ahead.

Much of this information isn't new. The portal builds on years of scientific knowledge and investments in geospatial platforms, but in the past these resources have been difficult to find and understand

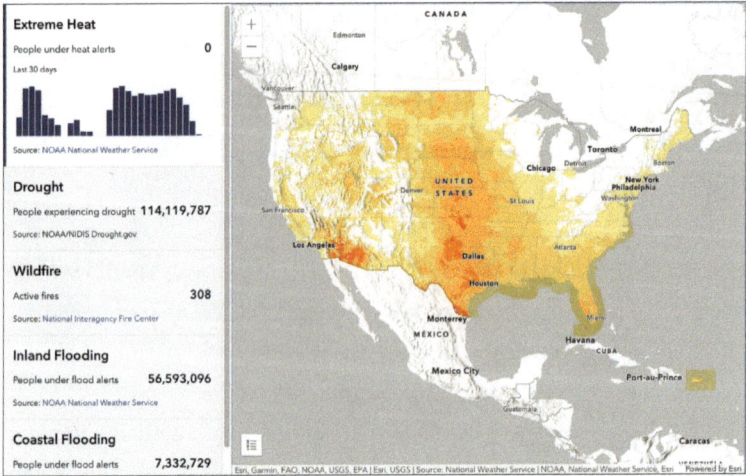

Climate Mapping for Resilience and Adaptation, or CMRA, integrates information from across the federal government to help people consider their local exposure to climate-related hazards, such as in this map of extreme heat.

because they were not integrated to provide a more complete picture. GIS technology plays a leading role, integrating information to help us understand our complex problems in the crucial context of location.

This portal brings the data together, for everyone to see—on maps, in charts, and in reports. Environmental data, as well as the social and economic factors that shape how well we can bounce back from climate-related hazards, can be explored by anyone, including city planners, resilience officers, transportation planners, tribal leaders, and residents.

## Visualizing our risks

Typing an address or picking a point on a map in the CMRA Assessment Tool reveals future climate projections related to extreme heat, drought, inland flooding, coastal flooding, and wildfire. The results show projections for the early years of the 21st century (2015–2044) and the middle (2035–2064) and late (2070–2099) years of the century, based on two climate change scenarios: one in which we reduce global emissions of heat-trapping gases to zero by about 2040, and one in which emissions increase through 2100. Among the many indicators are projections related to extreme rain or heat, number of consecutive dry days, and percentage of coastal counties impacted by global sea level rise.

Extreme heat can be judged based on the annual number of days exceeding 105 degrees Fahrenheit. In the census tract that includes the Esri headquarters in Redlands, California, for example, the results reveal possibly 22–42 days of extreme heat per year by the late century (compared with fewer than 4 days in 1990, the year used as a baseline for emissions calculations by many international organizations). In Washington, DC, the number of days with temperatures exceeding 95 degrees Fahrenheit grows from nearly 19 days per year in the early part of the century to as many as 64 days by the last years of the century.

Although climate action strategies have become commonplace in some larger metropolitan areas, including Boston, Miami, and Los Angeles, a data-driven approach may have been out of reach for others. The CMRA Portal aims to fill knowledge gaps—helping communities identify climate threats so they can prioritize resilience-building actions and discover programs offering funding to make solutions happen.

A state, local, or tribal leader could use the tool to better understand how temperature, precipitation, and flooding conditions are projected to change locally and generate straightforward hazard reports that can be incorporated into strategies for future projects, a climate action plan, or to support a data-driven proposal seeking funding. In addition to offering climate data, the tool also shows

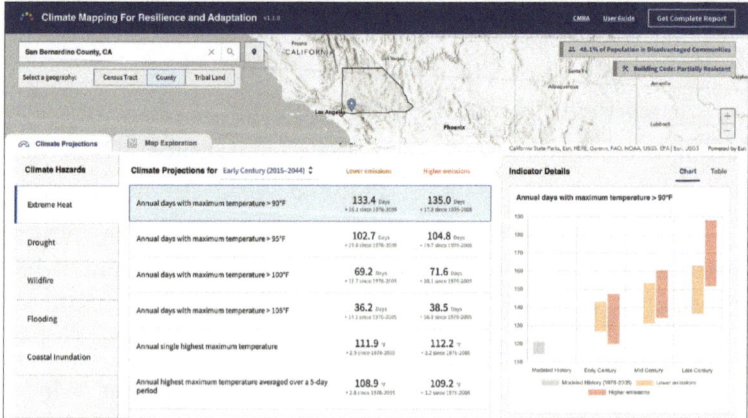

The CMRA tool gives users a national view as well as the ability to study local details, such as this analysis of extreme heat in Redlands, California.

areas designated as disadvantaged communities—based on an environmental justice score—making it eligible for prioritized funding. The aim is for equity to be a priority when government leaders design and implement resilience measures. The White House's Justice40 initiative ensures 40 percent of an eligible program's federal funding goes to benefit disadvantaged communities that are "marginalized, underserved, and overburdened by pollution."

A grant officer deciding who gets funding could use the tool to review proposed projects by their locations, ensuring that the funds go to projects that are the most needed now and that address a community's resilience for future generations. The officer could also help ensure that funding awards are distributed equitably by quickly verifying priority eligibility funding through the Justice40 initiative.

Concerned and curious residents could learn more about the climate-related hazards that may impact them or their neighbors, browse maps in the assessment tool, and create a report to share with others to inspire action.

## Providing evidence to back funding requests

In addition to information about projected climate conditions in the future, the portal highlights links to federal funding resources, federal climate policies, and proven solutions from other communities. The linked US Climate Resilience Toolkit offers videos and stories about what others are doing. For instance, one case study shows how a tribe in northern Wisconsin is replanting its forests to adapt to future conditions, and another shows how cities in Minnesota have modified their transportation system and extended the hours that cooling centers are open to ensure that people who need relief from the heat can get it.

Technologies such as the CMRA Portal enable dynamic collaboration—gathering information from across multiple agencies to enable better decisions, expose solutions that are working for others, and create a means for supporting evidence-based funding requests for all communities.

## Open data extends the value

The CMRA Portal's collection of relevant climate data supports decision-making and includes open data that can be accessed from the portal and combined with local GIS data or incorporated into assessment tools. The curated content and open data services invites users to configure new tools and maps that address local concerns.

As David J. Hayes, special assistant to President Biden for climate policy, said during the annual Esri User Conference in July 2022, "There is no more important service that the federal government can provide to all Americans right now than completely up-to-date information on the climate impacts, on a geographic basis, that are hitting our communities, causing the loss of lives and loss of livelihoods."

Today, we have better data, better tools, and even more appetite for climate information than in years past.

The portal can engage everyone and get leaders started with data-informed decision-making, leading to impactful interventions. Maps help people communicate, navigate, and see beyond where they stand, answering the question of what or who is on the other side of that mountain range, across that sea, or beyond the horizon.

Today's high-tech maps illustrate the conditions we face now and those we'll confront in the future. This portal has the potential to inspire people to encounter and address challenges collectively in hopes of a better future.

A version of this story by Patricia Cummens titled "White House Portal Helps Communities Assess Exposure to Climate Hazards" originally appeared on the *Esri Blog* on September 16, 2022.

# MAPPING PRIME RENEWABLE ENERGY SITES

### Kentucky Energy and Environment Cabinet

THE STATE OF KENTUCKY IS TRANSITIONING FROM COAL powerhouse to being a compelling locale for renewable energy generation. Once the leading producer of coal in the United States and still one of the top three coal-producing states, Kentucky envisions a future powered by alternative energy sources—hydropower, biomass, and solar.

Developers are looking for solar power locations as Kentucky faces declining coal use and production along with increasing interest from corporate buyers for renewable resources. With its strong infrastructure and available space, including previously used mine lands, Kentucky bills itself as an ideal choice for solar production sites. But where do they put them?

"For several years, as solar has come down in cost, it's become more of an option here in Kentucky," said Kenya Stump, executive director, Kentucky Office of Energy Policy. "We had a lot of questions from people: 'It seems like our mine lands would be great for solar,' or 'I don't know why we don't put solar on our mine lands.' And to me, that was always a geospatial question. Where should new sites go?"

Until a few years ago, the Office of Energy Policy had no mechanism to receive and respond to site inquiries, so developers often selected sites with little to no input from the state. "Solar is so new in Kentucky that we had solar developers making siting decisions, but we didn't understand why they were choosing their locations," Stump said. "We would just get notified that a new solar project was going here or going there."

Staff from the Kentucky Cabinet for Economic Development would pass along inquiries to colleagues at the Office of Energy Policy, who, initially, had no way to gauge suitability. Traditional industrial development sites weren't meant for solar, and siting characteristics such as topography, slope, or the presence of threatened species had to be considered. Teams at the Office of Energy Policy used their GIS technology to create the Solar Siting Potential in Kentucky platform, guiding solar developers to prime locations.

## A smarter solution for site selection

With input from the state's GIS experts and Esri, the Kentucky Energy and Environment Cabinet applied GIS to conduct site suitability analysis on land parcels available for development. The analysis evaluates sites based on criteria impacting the technical feasibility of construction, and then gives each parcel a score.

Stump's team collaborated with mining consultants and solar developers to collect relevant data layers and determine the criteria needed to compare sites. Support provided by the Kentucky Geography Network (KyGeoNet), the geospatial data clearinghouse for the state, facilitated this collaboration. Melissa Miracle, a former IT consultant who was deeply involved in KyGeoNet, worked closely with the state's nature preserves to include data on local endangered species. "That was a tough one because we didn't have direct access to the raw data," said Miracle, who now works for Esri. "So we worked with the folks within the agency who handle the nature preserves to create the layers that we needed. It was important for them to be involved so the developers would know where they can and cannot build." Miracle also worked with the Natural Resources division to create a better understanding of the state's mines.

"That was a whole learning curve," Stump said. "The attributes, how they code things, understanding mining reclamation terminology—all that was huge."

Through this collaboration, the team addressed the multifaceted concerns of solar plant providers—including favorable slope, land classification (barren land, mixed forest, and cultivated crops), access to electric transmission lines, population density, proximity to the habitats of threatened species, and status as federal or protected lands.

## Communicating with communities

Another benefit of Solar Siting Potential is its transparency to communities that may be affected by possible development. "Our land-use planning is done at the local level. So this tool also helps our local communities understand the reason why developers are looking at their lands—whether it's because they have the right slope or they have access to transmission or other characteristics," Stump said. "It's still up to that community to decide what they want to do with the land."

The platform also helps community stakeholders determine which local land is being considered for solar projects. This information influences personal and local planning decisions, such as safeguarding land for specific uses or protecting views.

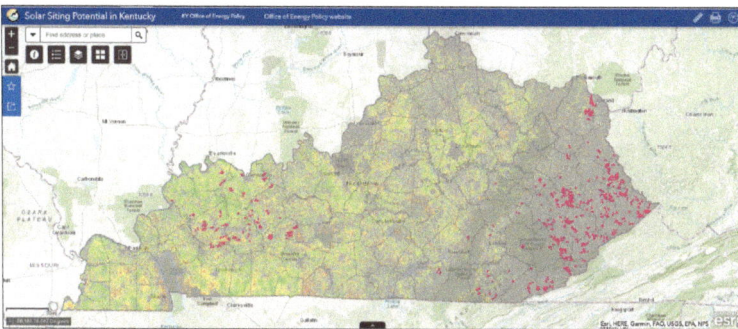

Interactive solar siting potential application created by the Kentucky Energy and Environment Cabinet.

"This tool couldn't have come at a better time because it shows stakeholders that we are actively trying to find places that were not prime farmland," Stump said. "We have a layer in the tool that you can turn on to show the land-use classifications, which can also inform the developer in the conversations with the community."

The program was planning 3D enhancements to the platform to make it easier for users to visualize the areas under consideration for development. "We're working on creating 3D map scenes in some of the prime areas for solar across the state," Miracle said. "This would allow developers to 'fly in' and see the slope, the terrain, a whole new view."

## New applications, new opportunities

Kentucky's success in analyzing solar siting potential has created opportunities for collaboration with other states. "Given the work our GIS group is doing, we're known nationwide as the office you go to if you want to learn how to get started with GIS," Stump said. "We're answering energy questions from other states that, at the heart of it, are geospatial questions. Other energy offices are beginning to see the light: that they need to know where things are going to occur, where they should occur, and where they could occur, before talking to stakeholders and discussing policies."

Kentucky's GIS experts envision using the technology to adapt the siting platform tool to incorporate environmental justice data.

"It's going to be another lens by which we look at everything we do, from emergency response to permitting to siting facilities and economic development projects," said Stump. "Right now, we're assessing where the datasets are. We know the Environmental Protection Agency (EPA) has the EJScreen environmental justice tool, the Census Bureau has Community Resilience Estimates, the Department of Energy has the LEAD tool to examine low-income energy affordability, but how do we bring them together? What does environmental justice for energy look like in Kentucky? That's a big 'where' question. We're really excited about where we can go."

A version of this story by Mike Bialousz titled "Finding a Home for Solar: Kentucky Maps Prime Renewable Energy Sites" originally appeared on the *Esri Blog* on July 22, 2021.

# EXPANDING COOLING CENTERS BASED ON EQUITY MAPS

## City of Los Angeles

IN LATE AUGUST 2022, A VICIOUS HEAT DOME SAT OVER LOS Angeles just when summer would typically fade into fall. For a week, temperatures climbed as high as 115 degrees in parts of Southern California and didn't cool off much at night. The extreme heat led to an increase in wildfire risk, power outages, and harm to human health and ecosystems.

While climate change affects everyone, there are disproportionate negative impacts on communities of color, older adults, young children, and outdoor workers. Predominantly Black and Latino lower-income neighborhoods in Los Angeles have less climate-adapted infrastructure than in wealthier communities. Having few trees and many buildings and roads creates an urban heat island effect. Concrete and asphalt store heat—warming the days and nights, leading to higher incidences of hospitalizations and premature deaths.

Extreme heat waves in California are lasting longer and are the deadliest climate threat residents face. There are efforts in low-income neighborhoods to add recreation center options and to plant trees to provide shade equity, but some of those projects, including the redesign of the Los Angeles River greenway, will take time to bring benefits.

In 2022, Los Angeles promoted Marta Segura to become the city's first chief heat officer. Applying GIS technology to map the city's areas most affected by heat and social vulnerability—a climate justice framework—helped Segura expand the city's services and deploy resources during extreme-heat events.

"LA now has six times the number of heat waves we once did,

with more exposure from heat-stagnated air pollution and less time for our bodies to recover," Segura said. "Some neighborhoods without trees, shade, and open space suffer four times the number of hospitalizations and premature deaths than areas with a greener, more climate-adapted infrastructure."

## Showing where residents can take respite from extreme heat

During the heat dome event, local emergency departments treated an additional 1,500 patients per day. According to the Center for Public Health and Disasters at the University of California, Los Angeles, 40 heat-related deaths were occurring daily by the fifth day of the heat wave.

"As the city's chief heat officer, I knew we had to act quickly to convey information about the city resources that could be a refuge from the heat to our communities," Segura said. With Segura's vision and support, the city launched Cool Spots LA, a web app showing residents where they can cool off during heat events. This is the first of many resources Segura is mandated to create as she develops the city's Heat Action Plan and Climate Vulnerability Assessment, which will prioritize extreme-heat response.

The app uses GIS technology in a single interactive directory to display more than 200 cooling centers. In the app, residents can input an address, cross streets, or a landmark or enable the app to use their current location to begin their search. In addition, the app includes a map with colorful icons to indicate specific types of cool spots such as hydration stations, recreation centers, public pools, and libraries.

To populate the map, Segura gathered data from the city's departments of emergency management, public works, water and power, and recreation and parks, as well as public libraries. Segura, in addition to being chief heat officer, leads the Climate Emergency

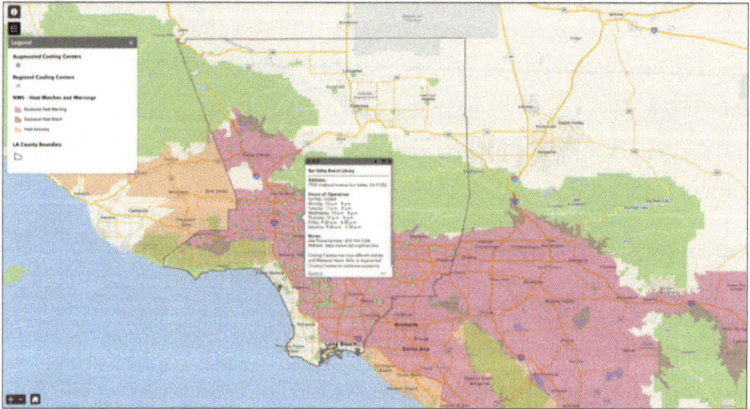

Visitors to the Cool Spots LA app can see the cooling centers across the region.

Mobilization Office. She coordinated support from Eva Pereira, the city's chief data officer on the mayor's data team, who used GIS to design and build the web app.

"This is how constituents expect city leaders to lead—we must cocreate and codesign services and projects—seamlessly," Segura said.

To promote a variety of locations and advance equity, the team gathered data on augmented cooling centers—found within 73 libraries, select recreation centers, and senior centers—and cooling spots such as shade structures, bus shelters, and hydration stations throughout Los Angeles to put in the Cooling Spots map.

## Visibility of cooling centers results in record relief

Seeing all the options on a map guided decision-making on where more were needed. The city added 16 augmented cooling centers in underserved areas from a previous count of 6.

"We had the highest cooling center attendance we have ever had, in part because of our app," Segura said. Calls to 311 and a social media campaign, #heatrelief4LA, promoted the app.

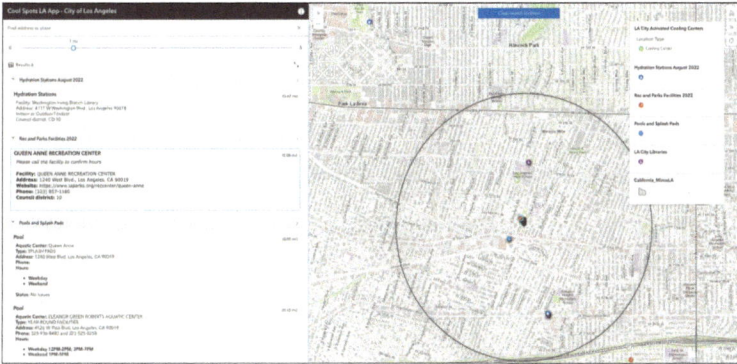

Zooming into the neighborhood level reveals a more detailed view to see centers by facility type.

The Cool Spots map will continue to grow and change over time as investments in green infrastructure expand. California governor Gavin Newsom also announced an $800 million heat action plan in 2022 to adapt and strengthen community resilience to heat threats.

In September 2022, Los Angeles Mayor Eric Garcetti announced an initiative to offer rebates for energy-efficient air conditioners—which is being informed by the Cool Spots LA app and the city's heat risk map. "The impacts of the climate emergency are on our doorstep, and as we continue our work to make Los Angeles a carbon-neutral city, we can't wait to bring solutions to people on the front lines of this crisis today," Garcetti said in a statement.

Los Angeles isn't the only city that has added an extreme heat officer. The Adrienne Arsht-Rockefeller Foundation Resilience Center's (Arsht-Rock) Extreme Heat Resilience Alliance catalyzed support for the concept and, worldwide as of late 2022, there were at least seven chief heat officers.

A version of this story by Patricia Cummens titled "LA's New Chief Heat Officer Expands Cooling Centers Based on Equity Maps" originally appeared on the *Esri Blog* on October 27, 2022.

# MAPS HELP ENSURE EQUITABLE DISTRIBUTION OF INFRASTRUCTURE FUNDS

## Montana Department of Natural Resources and Conservation

WHEN THE MONTANA DEPARTMENT OF NATURAL Resources and Conservation (DNRC) had a budget surplus, state officials analyzed maps and built online dashboards to share results with the public.

The $350 billion American Rescue Plan Act of 2021 (ARPA), an economic stimulus for state, local, territorial, and tribal governments, provided Montana $900 million to use strategically.

The Montana legislature decided to spend most of the money to upgrade the state's water and sewer systems. Staff at the DNRC were tasked with determining which communities to invest in. Montana DNRC invited city and county governments, state agencies, water and sewer associations, and conservation districts to submit proposals for projects to fund.

"They only had about a month to apply for this money, and we got over $900 million in grant requests," said Autumn Coleman, DNRC's resource development bureau chief. "For people to even apply for the money, they had to figure out how much they could get and how much in matching funds they needed to bring to the table to even be considered."

To allocate and disperse the funds, DNRC created two separate grant programs. One was to be divided among Montana's 56 counties, based on the same formula used to distribute money collected from the state's gasoline tax. This gas tax formula calculates a county's size, population, and miles of public roadways. The formula was fair but complex and difficult to parse for those unfamiliar with it. Adding further confusion, potential grantees were required to present matching funds.

An online dashboard and map proved crucial to the effort. Coleman and her team significantly reduced the complexity of the process by building a public-facing map and dashboard using GIS.

The visualization tools added a graphic element to what would otherwise be just facts and figures. Residents and grant applicants could more easily make a connection between the gas tax formula and its impact on funding decisions—and also monitor what decisions were being made as they began to appear on the map.

"It's just a really hard thing to explain to folks, so having this mapping tool made it easier for people to understand," Coleman said. "I think that's why spatial data works in this regard. We're trying to communicate with the public. We could share a giant 20-page table, but it just doesn't play as well as being able to look and see what you and your neighbors are getting."

## One big state, many small identities

The second ARPA grant program was purely competitive. Proposals were judged solely on merit. Geographic distribution was not officially a consideration, but that didn't mean it was going to be ignored.

Like all states, Montana has important political and demographic considerations, which are influenced by its size (third largest in the continental United States, twice the area of all New England states combined) and population (sixth lowest, 7 percent of New England's). Natural conflicts include urban centers versus rural areas, sparsely populated eastern Montana versus the more populous west, and longtime residents of modest means versus more affluent new arrivals. These conflicts play out against the backdrop of Montana's

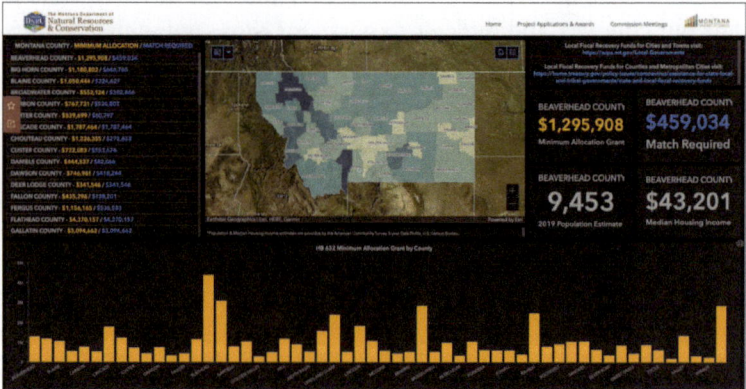

The dashboard of fiscal recovery fund spending on water and sewer infrastructure can be viewed by county or municipality.

rapid population growth, enough for the 2020 Census results to earn the state a new congressional seat. The fastest-growing area is Gallatin County, one of only two counties with a median family income greater than the US average. The upscale magazine *Travel + Leisure* recently called Bozeman, Gallatin's county seat, "one of America's coolest towns" and "a place of serendipities."

Despite its growth and coolness factor, Montana was recently ranked as the nation's seventh-worst state for infrastructure. According to Coleman, water and sewer funding is often viewed as an important dimension of fairness by the state's differing populations and places and their representatives.

"We're a very dispersed state, with a few major population centers and lots of rural areas," she said. "Legislators want to see ARPA funds spread evenly across Montana. They want to make sure that the smaller communities in their districts get a fair chance in all of this."

The map of city and town planned infrastructure projects shows a distribution across the state.

## Healthy competition

DNRC's map and dashboard of competitive grant applicants have served a dual purpose. As much as it promotes transparency by showing the public the geographic distribution of the grants, it also allows Coleman and her team to test various funding scenarios, which supports internal accountability.

"It was definitely a decision-making tool," Coleman said. "We could look at it and say, 'if we fund the top 28 projects, what does the map look like?' As we funded more projects, we could see the dots spread across the state.

"We could show that funds weren't just going to population centers," said Corey Richidt, a former GIS analyst and developer at DNRC. "We could even see the areas that are disproportionally affected by COVID."

For Brian Collins, GIS manager at DNRC, the maps serve as a manifestation of the agency's larger purpose—serving residents.

"Putting together this kind of information resource was a good reminder that we're in a public service profession," he said. "We're providing customer service at a very high level to people that need it right now. And it's very gratifying to put it out there this way."

A version of this story by Christopher Thomas titled "Maps Help Ensure Equitable Distribution of Infrastructure Funds in Montana" originally appeared on the *Esri Blog* on November 9, 2021.

# UNLOCKING A STATEWIDE UNDERSTANDING OF WATER QUALITY

## Kentucky Energy and Environment Cabinet, Division of Water

S CIENTISTS WITH THE KENTUCKY DIVISION OF WATER ARE looking for things the human eye can't see—such as bacteria and levels of pollutants—to determine whether a waterway meets minimum quality standards. But in some places, eyes can see clearly.

"I've been in streams in the middle of the Red River Gorge and it's just lovely," said Katie McKone, an environmental scientist with the Kentucky Division of Water. "Then I've been in other places I couldn't wait to get out of."

With nearly 92,000 river and stream miles, roughly 440,000 acres of lakes and reservoirs, and thousands of acres of springsheds (the area of land that contributes groundwater to a spring), Kentucky is rich with water resources.

To deal with the variability of water quality and report accurate findings to the US Environmental Protection Agency (EPA), McKone and her colleagues include GIS in their work to record, visualize, monitor contamination levels, and communicate results.

## Science and reporting

The US Clean Water Act requires states to report current water quality conditions every two years to the EPA. In 2022, in addition to the traditional integrated report, Kentucky developed an ArcGIS Hub℠ site with interactive maps and dashboards showing assessment results and impaired waters, while also providing explanations for concepts central to assessment, such as designated uses, reporting categories, monitoring programs, and assessment methods.

"On a map, you can look at the state and zoom in and ask an almost infinite number of questions," said Lara Panayotoff, supervisor of the total maximum daily load (TMDL) program at the Kentucky Division of Water. "You couldn't cover a millionth of what we cover in a static report."

For each assessed body of water, the maps and dashboards show what designated uses (such as swimming and fishing) have been assessed and whether it meets particular water quality standards. There are also links to TMDL reports, assessment summaries, and implementation summaries, so that all available resources for a waterbody are consolidated.

"You don't want to get it wrong," McKone said. "You don't want to call it impaired if it's not, and you don't want to say it's meeting the standards when it might be impaired. You have to have the data to support your decisions, and GIS is a key tool in the entire process."

With GIS, the team of water quality assessors can see the contours of the land in 3D—the ridges that mark the extent of every basin and how the water flows across the landscape. Mapping technology also lets them review land use to determine the location of possible pollution sources, such as farms, coal mines, manufacturing, and wastewater treatment facilities. Maps showing layers of data about the land and its uses guide the assessors' work and investigations.

## Peeling back the data layers

From the perspective of those reviewing the report, an interactive map, rather than pages of spreadsheets, turned out to be a much friendlier way to consume and understand an impressive amount of information.

"Just meeting the requirement of the water quality report yields a 100-to-200-page document," McKone said.

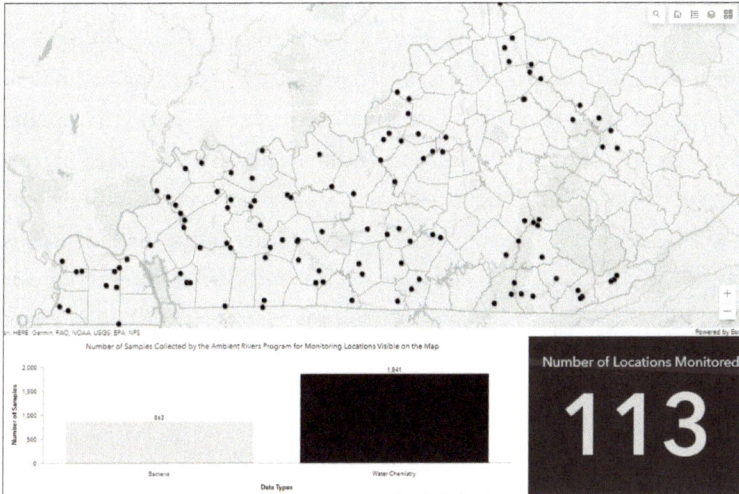

A dashboard about monitoring programs was developed to display where samples were collected, of what type, and how many from the time frame the report covers.

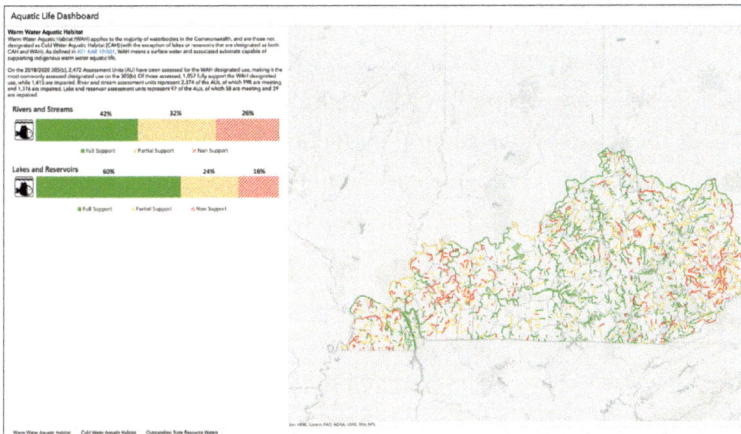

The Kentucky Division of Water's report to Congress contains many dashboards that detail quality measures, including this one that shows the health of aquatic life.

By building the report into an interactive website—integrated with narrative, maps, images, and dashboards—McKone had hoped to make it more accessible and transparent. "The report is an important resource where major statewide results are discussed, but it can't present a lot of detail about a given question someone might have," she said. "But the underlying information in the Hub site is different...it's totally cliché, but it's an onion. The interactive format allows a user to really peel back layers to get to what interests them most."

## Keeping everyone engaged

For the Kentucky water quality team and outside data contributors, the maps support ongoing work. Field biologists are endlessly gathering measurements such as water temperature and hardness, pH, and dissolved oxygen levels; then analyzing samples to detect metals like iron or manganese, nutrients including phosphorus and nitrogen, pesticides, herbicides, and bacteria.

With so much ground to cover across the state, the maps help everyone see where data has been collected and what the data says about the state's water quality. Individual businesses that are regulated now have easy access to this information, as do residents interested in seeing the quality of their local waterways.

Residents often reach out to the Division of Water when they see something in the water that concerns them, or when they just want to know about the health of their favorite creek or lake. "We'd be on the phone with somebody trying to talk through their questions, using spreadsheets and fairly coarse maps from the report. It was just difficult to communicate with each other," Panayotoff said. "Now, we can have them visit the interactive map, and we can walk through it together."

Putting the data on an interactive map has helped communicate water quality issues and successes, and has allowed the division to consolidate resources, educational materials, and program information in one place. While the water quality team knows that not all waterways will become the Red River Gorge, the hope is that all will meet water quality standards.

"We are public servants, and we need to be able to explain our programs so they can be understood by those we regulate and by the public, whose waters we are charged with protecting," McKone said. "By engaging more people, the hope is to increase awareness about water quality issues, and the work that is being done to address those issues, throughout Kentucky." Having more people engaged means having more partners in protecting and improving Kentucky's water quality.

A version of this story by Sunny Fleming and Christa Campbell titled "Unlocking a Statewide Understanding of Water Quality in Kentucky" originally appeared on the Esri Blog on July 19, 2022.

# MAPPING TO PROTECT NATURAL HERITAGE AND BIODIVERSITY

## South Carolina Natural Heritage Program

T O UNDERSTAND THE COMPLEX VARIABLES AND CHANGES in natural cycles that impact threatened species, conservationists are using sophisticated mapping techniques and GIS technology.

Consider the gopher tortoise and its unique relationship with forest fires.

The gopher tortoise is resilient to the occasional forest fire because the long tunnels it digs and inhabits 10 feet underground are fortified enough to withstand the smoke, flames, and burning debris.

Maintaining this subterranean lifestyle requires very specific habitat conditions. The sandy soils of the pine savannas found throughout the southeastern United States are particularly welcoming.

Over time, as the pine trees grow, the forest density threatens the vegetation the tortoises eat. Before major human contact, naturally occurring forest fires—often caused by lightning strikes—would help preserve this understory vegetation.

The modern fragmentation of the forests—through the introduction of roads, homes, croplands, and cities—has disrupted this cycle. Without fires, the forested areas that remain have become overgrown.

The decline of fire-adapted forest ecosystems is not the only factor that threatens the gopher tortoise, but it underlines the complexity of the fight to protect the endangered creature. The South Carolina Natural Heritage Program is arguably their most powerful ally.

## Protecting South Carolina heritage

The Natural Heritage Program—and its Cultural Heritage counterpart, dedicated to preserving land with historical and cultural relevance—together form the South Carolina Heritage Trust Program, a section within the state's Department of Natural Resources (SCDNR).

The work of the Heritage Trust began in the mid-1970s. Shortly thereafter, when the project was transferred to SCDNR, the state agency became the first state in the country dedicated to protecting land with abundant natural or cultural significance.

From the start, the work relied on one of the earliest desktop GIS software programs. As GIS software grew more sophisticated, Heritage Trust adopted an ArcGIS Enterprise approach. The technology became central to many of the program's objectives, including mapping habitat loss and picking areas the state could purchase for permanently protected heritage preserves.

With GIS, conservationists can analyze how and where to develop land in ways that protect biodiversity.

## Shortening the "long, drawn-out process"

The Natural Heritage Program recently expanded its GIS to facilitate better communication with stakeholders, including private

An interactive form with a map viewer helps developers submit plans for threatened or endangered species review.

developers, scientists, and the public. Communication and work-flows have always been challenges, especially regarding due diligence for developers proposing new projects. Any development that receives federal government financing or permitting must review the presence of threatened or endangered species within the proposed project footprint.

Until recently, developers would submit requests with varying degrees of specificity. Some contained only geographic coordinates, others included maps, and some were just general descriptions of the area. The office would check requests against its database of protected flora and fauna, struggling to get through around 200 requests per year.

"It was a long, drawn-out process, and it wasn't even our full-time job," said Joe Lemeris, the program's GIS and data manager.

The inefficiency extended to the workflow used by scientists who gathered data from the field. They recorded their observations onto

A sample South Carolina Department of Natural Resources report details the presence of endangered species and shares best practices to minimize impacts.

Excel spreadsheets or paper forms. At some later date, this information would become part of the program's database. The office also gathered data from other civic and private partners, including the US Fish and Wildlife Service, which could arrive in various formats, adding more process bottlenecks.

"The problem with the way it was done before is that the consultants could say they'd done due diligence, and that we'd shown them everything we had on a species, but then more data could pop up after we'd shared what we had," Lemeris said.

## Moving beyond static data for solar power

A few years ago, a proposed solar project northwest of Beaufort, South Carolina, spotlighted the outdated system. The developers consulted information previously obtained from the Natural Heritage Program database, unaware that data gathered more recently showed that current project plans threatened the gopher tortoise's habitat.

The problem, as Lemeris saw it, was that everyone was dealing with "static data"—information frozen in time as the world moved forward. What the program needed was a dynamic system that would reflect the most recent findings.

The solution was to take advantage of recent developments in GIS that prioritize the storage and flow of geographic information as well as its visualization. Data is now added and accessed through the same GIS-enabled portal. "It's essentially just a web app, and we added a custom-designed reporting tool," Lemeris said.

Scientists conducting field surveys input their data into ArcGIS Survey123 forms connected to the database.

"They have access to the data being collected in a way that goes beyond just writing it down in field notes and then having to transcribe it into a spreadsheet," said Tanner Arrington, GIS manager for

SCDNR. "They gain a locational context that they didn't have when the data was just tabular. With location attached to it, they can see patterns that were unknown before."

## Protecting species from foot traffic

The GIS data can be presented with varying degrees of specificity, depending on who wants to see it. For certain species, the Natural Heritage Program prefers not to share its survey data in granular detail, opting instead for a more generalized view.

The black rail bird merits this opaque approach. This bird's tiny stature restricts it to wetland areas with less than an inch of water. Unlike most birds, they spend more time running through marshes than flying through the air.

For many years, no black rails were sighted in South Carolina, and the bird remains on the threatened list under the Endangered Species Act. Recently, birdwatchers have logged black rail sightings in the state. While this is good news, birder enthusiasm can present a problem.

If the exact location of appearances by the black rail are revealed, the influx of birders may damage those spots. On the Natural Heritage Program's maps, only those with special permissions—scientists, mostly—can study exactly where rails have appeared. Others, including developers, will see general areas marked off as the bird's habitat.

## Starting "a ripple effect" with enterprise GIS

The success of the South Carolina Natural Heritage Program's portal has already paid dividends in increased efficiency. Lemeris estimates the tool has quadrupled the number of requests the office can process. The ease of use also encourages more developers to submit requests for data.

"I think there were a lot of people who weren't submitting projects to us because it was slow, and every day is more money," Lemeris said, "and also because it wasn't made very easy for them."

Other offices within SCDNR, as well as other state natural heritage programs, have also taken notice. "There's been a ripple effect," Arrington said. "The original understanding of the technology was through desktop GIS. But now that more people have seen what enterprise GIS is capable of, it's opened new opportunities within the agency. They're saying, 'Well, if it works for the Heritage Program, why can't we give it a try?' They've seen the benefits firsthand, and that's very powerful."

A version of this story by Mike Bialousz titled "Mapping to Protect Natural Heritage and Biodiversity in South Carolina" originally appeared on the *Esri Blog* on August 4, 2021.

# KEEPING DRINKING WATER SAFE

## Massachusetts Department of Conservation and Recreation

T HE QUABBIN RESERVOIR IS ONE OF ONLY FIVE MAJOR unfiltered drinking water supplies in the United States. It is part of the system that supplies water for more than 2.5 million residents in the Greater Boston area.

While boat-based fishing is permitted in certain areas of the reservoir, the potential for those boats to introduce aquatic invasive species into the water supply is a risk. That is why the Division of Water Supply Protection (DWSP), inside the Massachusetts Department of Conservation and Recreation's Office of Watershed Management, takes pains to protect the integrity of the Quabbin Reservoir and ensure that the only boats allowed are contaminant-free.

It does this via a boat seal program, whereby boaters wishing to use the reservoir must have their boats inspected, decontaminated, and sealed to their trailers by an unbreakable piece of wire with a special, numbered tag that DWSP tracks in a database. Only boats that pass the inspection are given a seal signifying that they are contaminant-free, which allows them to launch on the reservoir. And when these boats return to land, they are given new boat seals that tag them to their trailers. Boats without intact seals are prohibited from launching.

The boat inspection and decontamination program has historically relied on a combination of manual and paper-based workflows for scheduling inspections, maintaining seal data, and monitoring inspection activities. With about 200 boat inspections to perform each year, sustaining the various workflows throughout the entire process became inefficient.

"There was a huge input of staff time—over 700 hours annually—on translating paper-based information into other systems," said Erica Tefft, DWSP's watershed GIS coordinator. "It was, frankly, an expensive waste of people's time that they could have used to do other things."

The program's management team sought a modern, digital solution for managing the boat seal program. They wanted to streamline the data, as well as all the manual processes DWSP's environmental quality section was using to administer the master boater database.

After attending the 2018 Esri User Conference and regional user group meetings, Tefft learned about ArcGIS Survey123, a form-centric solution for creating surveys, collecting answers, and analyzing results. Survey123 was perfectly suited to capturing the tabular type of data required for boat inspections. In addition, the Survey123 Connector for Microsoft Flow would allow surveys to be configured so that when submitted, a trigger would initiate an automated action, such as sending specific data to the Google Calendar and Google Sheets DWSP uses.

For the boat seal program, Tefft designed two Survey123 surveys that together replaced the original manual processes.

The first survey Tefft built was the Quabbin Boat Seal Appointment and Decontamination Survey. Staff at the Quabbin Reservoir Visitors Center use it on desktop computers to set boat inspection and decontamination appointments, while Quabbin Watershed rangers use it on iPads to record critical details, such as regulatory requirements for each boat and motor, during the on-site appointments.

Next, Tefft built the Quabbin Boat Seal Program—In/Out Survey. Boat Launch Area (BLA) staff use this survey to check in boaters who have an inspection seal and sign them out with a new seal when they leave the water.

The two surveys capture all the data DWSP needs to ensure that the Quabbin Reservoir remains contaminant-free, including each boat owner's name and contact information, the boat registration number, the date of the decontamination, and the date and time of every launch.

To get the survey-captured data to automatically flow into other systems, Tefft used the Survey123 Connector for Microsoft Flow. From the Quabbin Boat Seal Appointment and Decontamination Survey, the Convert Time Zones action is triggered to automatically detect the appointment date and time, adjust to the correct time zone, and create an end time to reflect a 20-minute duration. The Create an Event Flow is automatically triggered as well, which makes a Google Calendar event. Now, staff use the Google Calendar to see all

When Boat Launch Area (BLA) staff submit a Quabbin Boat Seal Program—In/Out Survey, it automatically triggers the Tag_InOut_Flow in Microsoft Flow, which sends the data to a master spreadsheet.

the appointments for scheduled decontamination days, which eliminates the need to print out each day's itinerary.

When BLA staff submit a Quabbin Boat Seal Program—In/Out Survey, it automatically triggers the Tag_InOut_Flow in Microsoft Flow. This sends the data to a master Google Sheet that BLA employees, staff in the environmental quality section, and Quabbin Watershed rangers use to monitor boat activity on the reservoir for any given day. Having this data available in real time allows watershed rangers to know who is on or off the water at all times, satisfying a long-standing need.

Through Flow actions, the Download Survey123 Data Python script, and an R script, all the data from both surveys is automatically pulled into the master boater database, a repository that contains records for more than 2,300 boats. Having this information automatically entered into the database ensures better data quality and near-real-time information. Additionally, the new workflows for digitally collecting the data with surveys and automating the flow of data throughout DWSP's systems reduce the annual number of labor hours by 72 percent.

Using Survey123 and the Survey123 Connector for Microsoft Flow to digitally transform the manual workflows has enabled DWSP to realize a 137 percent initial-year return on investment in labor costs for the boat inspection program. All stakeholders now benefit from having accurate, complete, and up-to-date boat and boater data readily available at any time.

"Overall, using these surveys with the Survey123 Connector for Microsoft Flow is helping improve the quality of the boat/boater database to ensure the Quabbin Reservoir's water quality is being maintained at the highest possible level," concluded Tefft.

A version of this story titled "Massachusetts Keeps Drinking Water Safe" originally appeared in the Winter 2019 issue of ArcNews.

# NEXT STEPS

## The geographic approach to environmental management

AS COMMUNITIES ADDRESS THE IMPACTS OF CLIMATE change and seek to implement effective sustainable and resilient policies, environmental agencies face greater demands and expectations. Leading environmental agencies deploy GIS to promote equitable access to clean land, air, and water; implement effective natural resource management; and improve the quality of life with outdoor recreation. Using a geographic approach allows environmental and natural resource agencies to understand and communicate the relationship between social, economic, and environmental interactions. These new insights inform better decisions and lead to sustainable outcomes for communities, economies, and the environment.

Environmental agencies can take steps to modernize and expand the value of GIS across their organizations. In these next steps, you'll find suggestions and resources to help your organization realize the full potential of implementing the geographic approach to all aspects of your work and mission.

### Identify foundational data

You can gather and map foundational data in your area of interest. For many organizations, this process often means aggregating a hybrid of third-party datasets and geoenabling internal authoritative information. Organizations can also take advantage of ArcGIS Living Atlas of the World for a variety of relevant data layers, applications, pretrained deep-learning models, and more. For example, organizations can search ArcGIS Living Atlas for these and other kinds of foundational data:

- Soils and geology

- Weather and climate

- Streams and wetlands

- Near-real-time environmental sensors (streams, air quality, etc.)

- Biodiversity

- Administrative boundaries

- Demographic datasets

- Ecological systems

To learn more about live feeds, visit ArcGIS Living Atlas.

## Identify data gaps

Once existing foundational data is identified and organized, you can assess what else your organization may need. Environmental organizations often have workflows that result in an output of authoritative datasets. It's important to identify what those outputs may be, what roles within your organization may be responsible for the workflows, and the level of effort required to integrate them into your GIS. Often, it helps to use a graph to visualize the business process and its priority against its level of effort to help organizations prioritize their geospatial transformation.

## Create and share maps

Once your organization has the data sources it needs, you can more easily create and share maps and applications with your organization, stakeholders, and the public.

- **Real-time dashboards tracking environmental assets:** Keep your organization, stakeholders, and the public current on the real-time status of air quality, water quality, and other metrics to keep everyone informed and safe.

- **Storytelling maps to interpret and educate:** Stories created using ArcGIS StoryMaps help an audience engage with a specific topic, including environmental hazards, habitats, species, and more.

- **Organize volunteer initiatives with ArcGIS Hub:** Organize and engage with volunteers to conduct river or beach sweeps, invasive species days, environmental monitoring, and more.

- **Gather information using ArcGIS Survey123:** Provide better customer service with intuitive geoenabled forms so stakeholders can submit project proposals, apply for permits, report wildlife sightings, and more. Field staff and contractors can use these forms offline to conduct field inspections.

## Follow best practices

Ensure that your maps and apps are ready to scale and perform well at all times.

## Learn by doing

Hands-on learning will strengthen your understanding of GIS and how to use it to address environmental management. ArcGIS tutorials are a collection of free story-driven exercises allowing you to experience GIS applied to real-life problems. ArcGIS tutorials for environmental management include these topics:

- **Use species distribution patterns to assess protected areas.** Map protected areas and species information on the rarity of birds to understand how countries are protecting bird species with the highest rarity.

- **Configure apps for hikers.** Help park visitors search for nearby trails with a web app and a native app.

- **Use deep learning to assess palm tree health.** Identify trees on a plantation and measure their health using imagery.

- **Predict coral bleaching events in ArcGIS Online.** Explore the impact of sea surface temperature on coral bleaching with multidimensional data.

- **Monitor forest change over time.** Detect and analyze forest disturbance and recovery from a Landsat time series in the West Cascades ecoregion in Oregon.

- **Build a model to connect mountain lion habitat.** Find suitable corridors to connect dwindling mountain lion populations.

- **Model water quality using interpolation.** Use interpolation to analyze dissolved oxygen levels in Chesapeake Bay.

Try these and other free tutorials in the Esri Documentation Tutorial Gallery at https://learn.arcgis.com/en/gallery.

## Get there faster with ArcGIS Solutions

ArcGIS Solutions is a collection of focused maps and apps that help address challenges in your organization. They use your authoritative data and are designed to improve operations, provide new insight, and enhance services. Free with maintenance, ArcGIS Solutions can be deployed and ready to use with a single click or configured and extended to meet unique needs. They are industry-specific to address common workflows. You can draw from these Solution templates for environmental management:

- **Recreation Outreach.** The Recreation Outreach solution can be used to increase participation in outdoor activities and understand recreation license trends.

- **Park Infrastructure Management.** The Park Infrastructure Management solution can be used to inventory park assets, understand asset condition, and communicate changing asset conditions that impact park services.

- **Environmental Analysis.** The Environmental Analysis solution can be used to evaluate and understand the environmental impacts of proposed development projects and solicit feedback from stakeholders.

- **Community Science.** The Community Science solution can be used to solicit plant and animal observations from community scientists, manage each observation, and monitor community science programs.

You can view all these solution templates and more on the ArcGIS Solutions website.

## Learn more

For additional resources and links to live examples, visit the book web page:

go.esri.com/fasb-resources

# CONTRIBUTORS

Jim Baumann
Chris Chiappinelli
Keith Mann
Monica Pratt
Ben Smith
Citabria Stevens
Carla Wheeler

# ABOUT ESRI PRESS

ESRI PRESS IS AN AMERICAN BOOK PUBLISHER AND PART OF Esri, the global leader in geographic information system (GIS) software, location intelligence, and mapping. Since 1969, Esri has supported customers with geographic science and geospatial analytics, what we call The Science of Where®. We take a geographic approach to problem-solving, brought to life by modern GIS technology, and are committed to using science and technology to build a sustainable world.

At Esri Press, our mission is to inform, inspire, and teach professionals, students, educators, and the public about GIS by developing print and digital publications. Our goal is to increase the adoption of ArcGIS and to support the vision and brand of Esri. We strive to be the leader in publishing great GIS books, and we are dedicated to improving the work and lives of our global community of users, authors, and colleagues.

## Acquisitions

Stacy Krieg
Claudia Naber
Alycia Tornetta
Craig Carpenter
Jenefer Shute

## Editorial

Carolyn Schatz
Mark Henry
David Oberman

## Production

Monica McGregor
Victoria Roberts

## Sales & Marketing

Eric Kettunen
Sasha Gallardo
Beth Bauler

## Contributors

Christian Harder
Matt Artz
Keith Mann

## Business

Catherine Ortiz
Jon Carter
Jason Childs

# Related titles

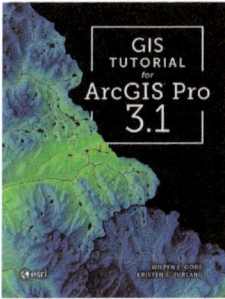

**GIST Tutorial for
ArcGIS Pro 3.1**

Wilpen L. Gorr and Kristen S. Kurland

9781589487390

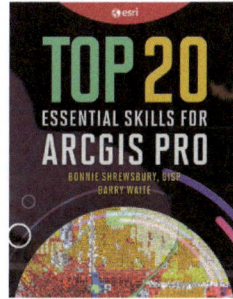

**Top 20 Essential Skills for
ArcGIS Pro**

Bonnie Shrewsbury and Barry Waite

9781589487505

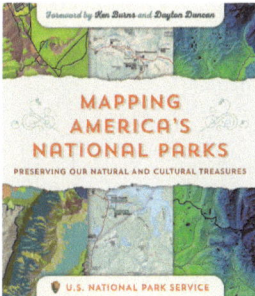

**Mapping America's National
Parks: Preserving Our Natural
and Cultural Treasures**

U.S. National Park Service

9781589485464

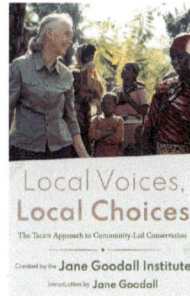

**Local Voices, Local Choices:
The Tacare Approach to
Community-Led Conservation**

Jane Goodall Institute

9781589486461

For information on Esri Press books, e-books,
and resources, visit our website at

## esripress.com.